Introduction to Pharmaceutical Biotechnology, Volume 3

Animal tissue culture and biopharmaceuticals

Introduction to Pharmaceutical Biotechnology, Volume 3

Animal tissue culture and biopharmaceuticals

Saurabh Bhatia and Satish Sardana
Amity Institute of Pharmacy, Amity University Haryana, Gurgaon, India

Tanveer Naved
Amity University, Noida, Uttar Pradesh, India

IOP Publishing, Bristol, UK

ISBN 978-0-7503-1347-6 (ebook)
ISBN 978-0-7503-1348-3 (print)
ISBN 978-0-7503-1349-0 (mobi)

DOI 10.1088/2053-2563/aafac0

Version: 20190301

IOP Expanding Physics
ISSN 2053-2563 (online)
ISSN 2054-7315 (print)

British Library Cataloguing-in-Publication Data: A catalogue record for this book is available from the British Library.

Published by IOP Publishing, wholly owned by The Institute of Physics, London

IOP Publishing, Temple Circus, Temple Way, Bristol, BS1 6HG, UK

US Office: IOP Publishing, Inc., 190 North Independence Mall West, Suite 601, Philadelphia, PA 19106, USA

Dedicated to my beloved parents, my wife (Mrs Sonali Bhatia Sachdeva) and my sweet little son (Daksh Bhatia).

Contents

Preface

Animal biotechnology covers several applications of animal biotechnology, especially biopharmaceuticals for animal or human use. This is the third volume of animal biotechnology which includes exhaustive information of animal tissue culture and biopharmaceuticals. This volume has been written with a view to providing background knowledge of the state of the art on the subject to date, and a practical review of the developments to date in animal tissue culture. The book covers several different facts of evolutional progress and achievements. Attention is focused on how animal cells respond against *in vitro* conditions. Emphasis is given more on cellular processes than molecular. With phenomenal progress in the application of animal tissue culture, various *in vitro* methods have been developed to define the correct requirements of successful culture. Some of the advancements in animal cell characterization include comparative genomic hybridization, epigenetic profiling, fluorescence in situ hybridization, karyotyping, single nucleotide polymorphism, pluripotency markers (proteins), stem cell arrays, flow cytometry etc. Other highlights pertaining to biopharmaceuticals are discussed in the final chapter, which gives an overview of the types, development and delivery of biopharmaceuticals, along with roles and responsibilities of pharmacists for biopharmaceuticals. It is hoped that graduate students, research workers and good commercial laboratories would find this book a useful adjunct in the understanding of intricacies associated with the culture.

Acknowledgements

This book is dedicated to the memory of my father in law Mr Ramesh Sachdeva, pictured. I also wish to thank my wife, Mrs Sonali Bhatia Sachdeva, who has stood by me through all my travails, my absences, my fits of pique and impatience. She gave me support and help, discussed ideas and prevented several wrong turns. Along with her, I want to acknowledge my son, Daksh Bhatia, as he is a great source of love and relief from scholarly endeavors. My special thanks go to my parents, Mr R L Bhatia and Mrs Vinod Bhatia and my brother Mr Sanjay Bhatia, whose support and blessings have enabled me to accomplish my research work successfully. I would also like to thank my sweetest mother in law, Mrs Nirmal Sachdeva, for her constant love and support. I am also grateful to Amity Institute of Pharmacy, Amity University Haryana for providing excellent facilities and a wonderful environment for the successful accomplishment of this book.

S Bhatia

Author biography

Saurabh Bhatia

Dr Saurabh Bhatia graduated from Kurushetra University, India, in 2007, followed by post graduation in 2009 from Bharati Vidyapeeth University, India. He successfully completed a PhD program in pharmaceutical technology at Jadavpur University, Kolkata, India (2015). Currently, he is working as an Associate Professor at Amity institute of Pharmacy, Amity University Haryana, Gurgaon, India. He has nine years of academic experience and his areas of interest include nanotechnology (drug delivery), biomaterials, natural products science, biotechnology (plant and animal), microbiology, modified drug delivery systems, analytical chemistry, parasitology (leishmaniasis) and marine science. He has promoted several marine algae and their derived polymers throughout India and has published more than five books and thirty articles in many areas of pharmaceutical science. He has participated in more than thirty national and international conferences.

IOP Publishing

Introduction to Pharmaceutical Biotechnology, Volume 3
Animal tissue culture and biopharmaceuticals
Saurabh Bhatia, Tanveer Naved and Satish Sardana

Chapter 1

Introduction to animal tissue culture science

1.1 Introduction

Animal tissue culture technology is now becoming a significant model for many scientists in various fields of biology and medicine. Despite the various developments in animal cell and tissue culture since the late 1800s, until the early 1950s progress in animal tissue culture was stalled due to the non-availability of a suitable cell line. In the early 1950s, for the first time, successful growth of cells derived from the cervical cancer of Mrs Henrietta Lacks was demonstrated. This breakthrough using Mrs Henrietta Lacks's cells in culture successfully transformed medical and biological research, allowing numerous cellular, molecular and therapeutic discoveries, including the breakthrough of the first effective polio vaccine [1, 2]. This culture is now called HeLa, on which there were more than 60 000 publications by 2017, and which has been involved in numerous Nobel prize-winning innovations [2–4].

Animal cell culture is a significant tool for biological research. The importance of cell culture technology in biological science was realized a long time ago. Earlier dedifferentiation based experiments of cells due to selective overgrowth of fibroblasts resulted in the enhancement of culture techniques. Animal cell culture involves isolation of cells from a tissue before establishing a culture in a suitable artificial environment. Initial isolation of the cells from the tissues can be achieved by disaggregation using enzymatic or mechanical methods. The source of the isolated cells is usually an *in vivo* environment, but sometimes cells are also derived from an existing cell line or cell strain. Animal cell culture offers suitable model systems for investigating the following factors:

- Drug screening and development.
- Mutagenesis and carcinogenesis.
- Normal physiology and biochemistry of cells.
- Potential effects of drugs and toxic compounds on the cells.

doi:10.1088/2053-2563/aafac0ch1

In addition, it also permits reliable and reproducible results, and is thus considered as a significant model system in cellular and molecular biology. Mammalian cell culture requires an optimal environment for growth. Environmental conditions are divided into nutritional requirements and physicochemical requirements. Nutritional requirements include a substrate or medium that provides support and essential nutrients such as amino acids, carbohydrates, vitamins, minerals, growth factors, hormones and gases (O_2, CO_2). All these factors control physical and chemical factors such as pH, osmotic pressure and temperature. In animal tissue culture the majority of cells are anchorage-dependent and therefore require a solid or semi-solid support in the form of a substrate (adherent or monolayer culture), whereas others can be cultured in the culture medium, called a suspension culture. Cell culture technologies have emerged as a tool to assess the efficacy and toxicity of new drugs, vaccines and biopharmaceuticals, and also play a major role in assisted reproductive technology. Animal cell culture is one of the more important and diverse techniques in current research streams. Animal, plant and microbial cells are always cultured in predetermined culture medium under controlled laboratory conditions. Animal cells are more complex than micro-organisms. Due to their genetic complexity it is difficult to determine the optimum nutrient requirements of animal cells cultured under *in vitro* conditions. Animal cells require additional nutrients compared to micro-organisms, and they usually grow only when attached to specially coated surfaces. Despite these challenges, different types of animal cells, including both undifferentiated and differentiated ones, can be cultured successfully.

1.2 Historical background

Tissue culture involves the *in vitro* maintenance and propagation of cells in optimal conditions. Culturing animal cells, tissue or organs in a controlled artificial environment is called animal tissue culture. The importance of animal tissue culture was initially realized during the development of the polio vaccine using primary monkey kidney cells (the polio vaccine was the first commercial product generated using mammalian cell cultures). These primary monkey kidney cells were associated with many disadvantages [5–8] such as:

- Chances of contamination with adventitious agents (risk of contamination by various monkey viruses is high).
- Most of the cells are anchorage-dependent and can be cultured efficiently only when they are attached to a solid or semi-solid substrate (obligatorily adherent cell growth).
- The cells are not well characterized for virus production.
- A scarcity of donor animals as they are on the verge of extinction.

The foundation of animal tissue culture can be considered to have occurred in 1880, when Arnold showed that leukocytes can divide outside the body [9]. Then, in the beginning of the 19th century, Jolly investigated the behavior of animal cells in serum lymph [9]. The development of animal tissue culture commenced after the breakthrough frog tissue culture technique, which was discovered by Harrison in

1907. Due to this effort Harrison is considered as the father of tissue culture. In his experiment he introduced tissue from frog embryos into frog lymph clots and showed that not only did the tissue survive, but nerve fibers grew out from the cells. During the mid-20th century, human diploid fibroblast cells were established by Hayflick and Moorhead [10]. They named this cell line MRC-5 (a cell line of fibroblasts derived from lung tissue). Later, Wiktor *et al* (1964) explored the utilization of this cell line in the production of rabies virus for vaccine production [11]. After a couple of years they suggested a large-scale production protocol along with a method for the assessment of purified rabies vaccine immunogenicity. During the same time, BHK-21 (C13) cells (baby hamster kidney cells) were established. These cells are susceptible to human adenovirus D, reovirus 3 and vesicular stomatitis virus. The commercial production of inactivated foot and mouth disease (a viral disease that causes sores in the mouth and a rash on the hands and feet of children) vaccine began using a suspension process [12]. Back in 1914, Losee and Ebeling [13] cultured the first cancer cells and after a few decades the first continuous rodent cell line was established by Earle (1943) [14]. In 1951, Gay established that human tumor cells can give rise to continuous cell lines. The cell line considered as the first human continuous cell line was derived from a cancer patient, Henrietta Lacks, as mentioned above, and HeLa cells are still used very widely. Continuous cell lines derived from human cancers are the most extensively used resource in the modern laboratory. The HeLa discovery was followed by FDA approval for the production of interferon from HeLa cell lines [15]. In addition to the progress in the field of cell culture, different media have been explored, which are typically based on specific cell nutritional requirements, such as serum-free media, starting with Ham's fully defined medium in 1965. In the 1970s, serum-free media were optimized by the addition of hormones and growth factors. Currently, thousands of cell lines are available and for the establishment and maintenance of these cell lines many media are available.

1.3 Types of cell cultures

Broadly, animal tissue culture can be divided into two categories:

- Cultures that allow cell–cell interactions and encourage communication or signaling between cells.
- Cultures in which cell–cell communication or interactions are lost or the signaling between them is missing.

The first category includes three different types of culture systems: organ cultures, histotypic cultures and organotypic cultures. The second category includes cultures in monolayers or as suspensions. Organ culture is a culture of native tissue that retains most of the *in vivo* histological characteristics, whereas culturing cells for their re-aggregation to yield tissue-like structure is known as histotypic culture. In histotypic cultures, individual cell lineages are initially derived from an organ and then cultured separately to high density in a 3D matrix to study interactions and signaling between homologous cells. In organ cultures, whole embryonic organs or small tissue fragments are cultured *in vitro* in such a manner that they retain their tissue architecture, i.e. the characteristic distribution of various cell types in the given organ.

In an organotypic culture, cells from different origins are mixed together in specific proportions and spatial relationships so as to re-form a component of an organ, i.e. the recombination of different cell types to yield a more defined tissue or organ. Some terms frequently used in animal tissue culture are as follows.

Cell culture. Cell culture is the process of removing cells from an animal or plant and their subsequent growth in an artificially controlled environment.

Primary cell culture. This is the first culture (a freshly isolated cell culture) or a culture which is directly obtained from animal or human tissue by enzymatic or mechanical methods. These cells are typically slow growing, heterogeneous and carry all the features of the tissue of their origin. The primary objective of this culture is to maintain the growth of cells on an appropriate substrate, available in the form of glass or plastic containers, under controlled environmental conditions. Since they are directly obtained from original tissue they have the same karyotype (number and appearance of chromosomes in the nucleus of a eukaryotic cell) as the original tissue. Once subcultured, primary cell cultures can gives rise to cell lines, which may either die after several subcultures (such cell lines are known as finite cell lines) or may continue to grow indefinitely (these are called continuous cell lines). Usually, normal tissues give rise to finite cell lines, whereas cancerous cells/tissue (typically aneuploid) give rise to continuous cell lines. Nevertheless, there are some exceptional examples of continuous cell lines which are derived from normal tissues and are themselves non-tumorigenic, e.g. MDCK dog kidney, fibroblast 3T3, etc. The evolution of continuous cell lines from primary cultures is assumed to involve mutation, which alters their properties compared to those of finite lines. Serial subculturing of cell lines over time can increase the chances of genotypic and phenotypic variation. Bioinformatic studies based on proteomic phenotypes discovered that the Hepa1–6 cell lines lacked mitochondria, reflecting a rearrangement of metabolic pathways in contrast to primary hepatocytes. With the emergence of newer technologies such as 3D culture, the use of primary cells is becoming increasingly prevalent and achieving improved results. Primary cells which are directly obtained from human or animal tissue using enzymatic or mechanical procedures can be classified into two types:

- *Anchorage-dependent or adherent cells.* Adherent cells are those cells which require attachment for growth and are also called anchorage-dependent cells. In other words, these cells are capable of attaching on the surface of the culture vessel. These types of cells are often derived from the tissues of organs, for example from the kidney, where the cells are immobile and embedded in connective tissue.
- *Anchorage-independent or suspension cells.* Suspension cells do not require attachment or any support for their growth and are also called anchorage-independent cells. All suspension cells are isolated from the blood system, for example white blood cell lymphocytes, and are suspended in plasma.

For several reasons cells obtained from primary cultures have a limited life span, i.e. the cells cannot be maintained indefinitely. An increase in cell numbers in a primary culture results in exhaustion of the substrate and nutrients, which can influence cellular activity and lead to the accumulation of high levels of toxic metabolites in the culture. This may ultimately result in the inhibition of cell growth. This stage is called the confluence stage (contact inhibition), when a secondary culture or a subculture needs to be established to ensure continuous cell growth.

Secondary cell culture. This simply refers to the first passaging of cells, a switch to a different kind of culture system, or the first culture obtained from a primary culture. This is usually carried out when cells in adherent cultures occupy all the available substrate or when cells in suspension cultures surpass the capacity of the medium to support further growth, and cell proliferation begins to decrease or ceases completely. So as to maintain optimal cell density for continued growth and to encourage further proliferation, the primary culture has to be subcultured. This process is known as secondary cell culture. Major differences between primary and secondary cell cultures are highlighted in table 1.1.

Table 1.1. Differences between primary and secondary cell cultures.

Primary cell culture	Secondary cell culture
Directly obtained from animal or plant tissue.	Originates from a primary cell culture.
Closely resembles the parental tissue.	Does not closely resemble the parental tissue.
The biological response of the cell may be closer to that in an *in vivo* environment.	The biological response of the cell differs from that an *in vivo* environment.
The first culture derived from original cells/tissue (from an *in vivo* environment).	Derived from an existing culture.
Cannot be transformed.	Can be transformed.
Less chance of mutation.	Can increase the chance of mutation or genetic alteration of primary cells.
Acquired through steps of rinsing, dissection, and mechanical or enzymatic disaggregation.	If the primary culture is an adherent culture, the first step is to detach cells from the attachment (the surface of the culture vessel) by mechanical or enzymatic means. Then, the cells have to be detached from each other to form a single-cell suspension.
Finite life span.	Prolongs the life span of cells. Periodic subculturing may produce immortal cells through transformation or genetic alteration of primary cells.
The risk of contamination is high. More difficult to maintain.	The risk of contamination is lower. Comparatively easy to maintain.

Cell line. Once a primary culture is subcultured or passaged it represents a cell line. A cell line that experiences indefinite growth of cells during subsequent subculturing is called a continuous cell line, whereas finite cell lines experience the death of cells after several subcultures.

Cell strain. A cell line is a permanently established cell culture which will proliferate forever if a suitable fresh medium is provided continuously, whereas cell strains have been adapted to culture but, unlike cell lines, have a finite division potential. A cell strain is obtained either from a primary culture or a cell line. This is done by selection or cloning of those particular cells having specific properties or characteristics (e.g. specific function or karyotype) which must be defined.

In summary, the first culture that is established from the *in vivo* environment is called the primary culture. This primary culture can be subcultured many times to develop cell lines. Cell lines are generally immortalized or transformed cells, i.e. cells that have lost control over division, because of mutations or genetic alterations, or because a primary cell was transfected with some genes that immortalized the cells. Most cell lines are tumorigenic as they originated from tumors. Cells derived from a primary cell line do not have this concern, however, it is challenging to maintain these cells. In usual practice, primary cell cultures require a nutrient medium containing a high amount of different amino acids, micronutrients and, occasionally, some types of hormones or growth factors. Primary cell cultures can be efficiently utilized up to a few passages, about two to four, afterwards their risk of contamination is higher than for cell lines. However, primary cell cultures have their own advantages. The biological response received from a primary culture will be closer to that in an *in vivo* environment than the response obtained from cell lines. From many years, several cell lines have been established and tested under different environmental conditions. This vast research has resulted in a good amount of data supporting the use of specific cell lines as models of primary cells. It has been suggested that cell lines that have been well tested under different conditions should be used instead of primary cultures, in the case that the latter are expensive.

1.4 Primary cell culture

As discussed above, the primary cell culture is the first culture of cells, tissues or organs derived directly from an organism; in other words it is the culture before the first subculture, whereas the cell line is for maintenance or propagation of a culture after subculture. There are certain techniques available for the development of primary cell cultures, such as:

- Mechanical disaggregation.
- Enzymatic disaggregation.
- Primary explant techniques.

1.4.1 Mechanical disaggregation

It is necessary to disaggregate soft tissues such as soft tumors. The mechanical approach involves slicing or harvesting tissue and subsequent harvesting of spill out cells. This can be achieved by sieving, syringing and pipetting. This procedure is inexpensive, rapid and simple, however, all these approaches involve the risk of cell damage, thus mechanical disaggregation is only used when the viability of the cells in the final yield is not very important.

1.4.2 Enzymatic disaggregation

This approach involves efficient disaggregation of cells with high yield by using enzymes such as trypsin, collagenase and others. Enzyme based disaggregation allows hydrolysis of fibrous connective tissue and the extracellular matrix. Currently, the enzymatic method is extensively used as it offers high recovery of cells without affecting the viability of cells.

1.4.2.1 Trypsin based disaggregation or trypsinization

This allows disaggregation of tissue using trypsin, usually crude trypsin because this trypsin contains other proteases. In addition, cells can tolerate crude trypsin well and the ultimate effect of crude trypsin can easily be neutralized by serum or trypsin inhibitor (supplementation of trypsin inhibitor is required in the case of serum-free media). Pure trypsin can also be utilized for disaggregation of cells, provided that it is less toxic and very specific in its action. An overview of primary cell culture development is shown in figure 1.1. Two common approaches, namely warm and cold trypsinization, are described in the following.

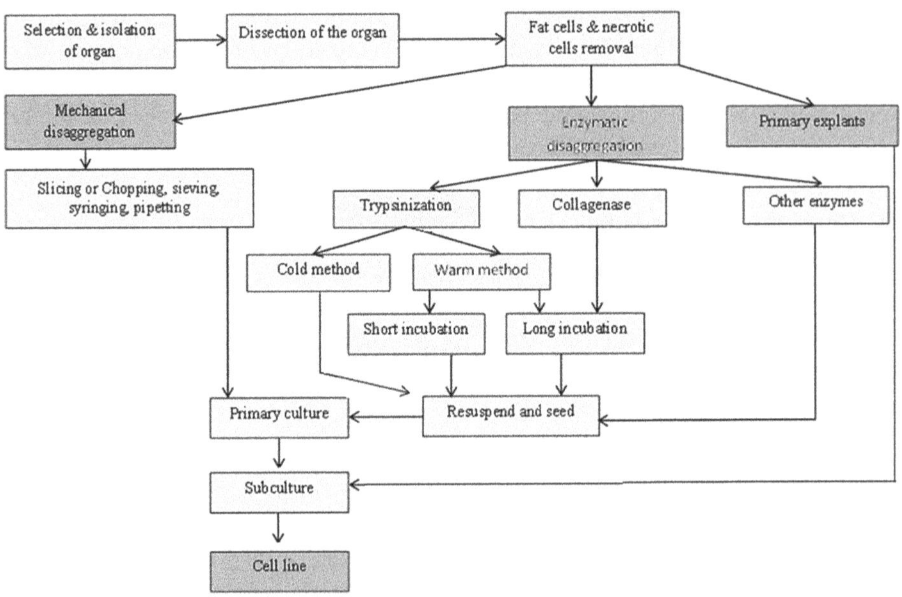

Figure 1.1. Alternative approaches for the preparation of primary cell cultures.

Warm trypsinization

This approach is extensively utilized for the disaggregation of cells. During the initial step, sliced tissue is washed with dissection basal salt solution and is subsequently transferred to a container of warm trypsin (37 °C). At regular intervals of 30 min the contents are stirred properly. Then, the supernatant having dissociated, the cells are separated to disperse in a suitable medium. Efficient dispersion of cells can be achieved by placing the container over ice.

Cold trypsinization

This method is also called trypsinization with cold pre-exposure. In this process the chance of cellular damage due to constant exposure to trypsin is reduced, which results in a high yield of viable cells with an improved survival rate for the cells (after 24 h of incubation). Since this method does not involve frequent stirring or centrifugation, it can be conveniently adopted in the research laboratory. During this process, after washing and chopping, tissue pieces are kept over ice in a vial and then subjected to treatment with cold trypsin for 6–24 h. Then, after the cold trypsin treatment the trypsin is removed and discarded. However, the tissue fragments still contain residual trypsin. These fragments are incubated at 37 °C (for 20–30 min) followed by repeated pipetting. This will encourage the dispersion of cells. The fully dispersed cells can be counted using a cell counter and properly diluted, and then further utilized.

Drawbacks of trypsin disaggregation

Trypsinization of cells can damage some cells, such as epithelial cells, and sometimes it is not effective for certain tissues, such as fibrous connective tissue, thus other enzymes are also recommended for dissociation of cells.

1.4.2.2 Collagenase based disaggregation

Collagenase is an enzyme which is responsible for the cleavage of peptide bonds in collagen. Collagen is a structural protein which is abundantly found in higher animals, mainly in the extracellular matrix of connective tissue and muscle. Collagenase, mainly crude collagenase, can be successfully used for the disaggregation of several tissues that may or may not be sensitive to trypsin. Purified collagenase has also been experimented with, but has shown poor results in comparison to crude collagenase. So far collagenase disaggregation has be carried out on several human tumors, epithelial tissues, the brain, lungs and other mammalian tissue. The combination of collagenase with hyaluronidase offers better results in disaggregating rat or rabbit liver, which can be achieved by perfusing the whole organ *in situ*. Several researchers have also utilized trypsin and collagenase in combination to dissociate cells to develop chick serum.

This process involves an initial transfer of the desired tissue into a basal salt solution which contains antibiotics. This is followed by washing with settling and then transfer into a medium containing collagenase. The solution is incubated for 1–5 days, followed by repeated pipetting for uniform dispersal of cells. Separation of

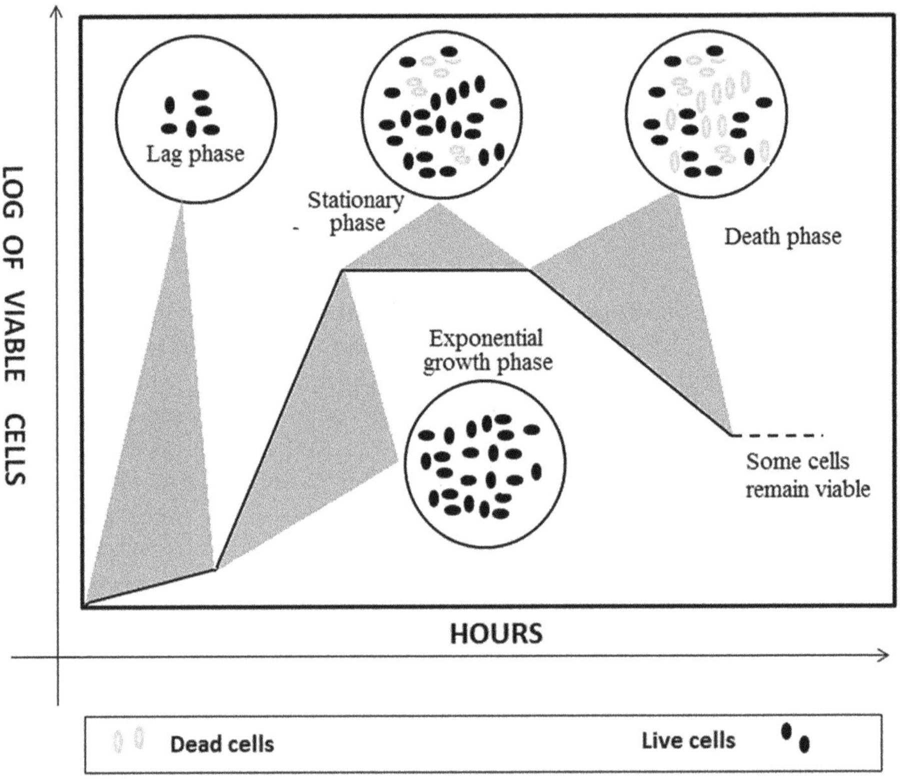

Figure 1.2. Standard growth curve of cells in a culture.

these dispersed cells is encouraged by keeping the solution in a stationary phase to further encourage the settling of cells, as shown in figure 1.2.

1.4.2.3 Other enzymes

In addition to the above mentioned enzymes, certain other enzymes such as bacterial proteases (e.g. dispase, pronase) have been tested, but unfortunately have not shown significant results. However, enzymes such as hyaluronidase and neuraminidase have received attention due to their significant results, and thus can potentially be utilized in conjugation with the enzymes discussed above.

1.4.3 Primary explant technique

In 1907 Harrison provided the first demonstration of the primary explant technique, which subsequently underwent many modifications. A simple protocol for the primary explant technique is represented in figure 1.1. As in the above procedures, in this process tissue is initially suspended in basal salt solution and then chopped properly and washed by settling. Tissue fragments are uniformly distributed over the growth surface. This is followed by the addition of a suitable medium and then incubation for 3–5 days. Old medium is replaced by fresh medium unless desired growth or considerable outgrowth of the cells is not achieved. Once optimum growth is achieved the explants are separated and transferred to new culture vessels which contain fresh medium.

This technique is mainly used for disaggregation of small quantities of tissue. Mechanical and enzymatic disaggregation are not suitable for small amounts of tissues, as there is a risk of cell damage which can ultimately affect cell viability. A major drawback of this technique is the poor adhesiveness of certain tissues on the growth surface (substrate material), which can create problems in the selection of cells for desirable outgrowth. However, this technique has been utilized frequently for culturing embryonic cells, in particular glial cells, fibroblasts, myoblasts and epithelial cells.

1.5 Segregation of non-viable cells from viable cells

After the development of a primary cell culture, it is essential to remove the non-viable cells from the disaggregated cells, which can be achieved by repeatedly changing the medium. Only a few will be left after dilution of the medium, and finally will gradually disappear when viable cells start proliferating. The alternative approach of centrifugation, mixing cells with ficoll and sodium metrizoate, can also be utilized to remove non-viable cells from the primary cell culture. Dead cells form a pellet at the bottom which can easily be removed from the solution.

1.6 Ethical issues in animal tissue culture

Animal tissue culture techniques involve the frequent utilization of animal or human tissues, which raises the need for safety and ethics guidelines for using animals in research, also known as medical ethics. Handling animals raises numerous issues that are typically not faced when using animal tissue. In addition to the consent of local ethical committees, the consent of the patient or his/her relatives is required to initiate research or to study a human sample in the form of fetal materials or biopsy samples. Samples collected from a human donor should be accompanied by a donor consent form in a prescribed format. When dealing with human tissue, the following issues should be considered [16]:

- The patient's or relatives' consent for using tissue for research purposes.
- Ownership of specimens, in particular cell lines and their derivatives, i.e. the recipient will not trade or transfer the cell lines and their derivatives.
- Consent for genetic modification, in particular in the case of cell lines.
- Patent or intellectual rights for the commercial use of cell lines.
- Guidelines should be refined to meet the requirements of the latest ongoing developments in animal tissue culture science. These guidelines are framed to offer knowledge to those new to the field and others involved in training and instruction, with the data required to improve their awareness of issues and to allow them to deal with them more efficiently. The primary areas of focus in guidelines are:
    i. Acquisition of cell line.
    ii. Authentication of cell line.

iii. Characterization of cell line.
iv. Cryopreservation of cell line.
v. Development of cell line.
vi. Instability of cell line.
vii. Legal and ethical requirements when deriving cell lines from human and animal tissues.
viii. Microbial contamination of cell line.
ix. Misidentification of cell line.
x. Selection and maintenance of equipment.
xi. Transfer of cell lines between laboratories.

Generally, when dealing with human tissue, the donor/relative is asked to sign a disclaimer statement in a prescribed format before tissue is collected. Doing this can reduce the chances of legal problems [16].

1.7 Safety considerations in animal tissue culture

Handling human tissue involves a high risk of exposure to various infections, thus it is essential to handle human material in a biohazard cabinet. Before their use, tissues must be screened properly for various infections such as hepatitis, tuberculosis and HIV. In addition, media, apparatus and glass wares should be properly sterilized (autoclaved) to considerably reduce the chances of spreading any infections.

1.8 Cell lines (first subculture or passage)

A cell line can be defined as a permanently established cell culture which will propagate forever, provided the continuous supply of suitable fresh medium and the availability of space for the cells to propagate. Thus, generally, a cell line can be defined as the propagation of a culture after the first subculture. In other words, when primary culture is subcultured it results in the development of a cell line. Cell lines differ from cell strains in that they become immortalized. A cell line contains several cell lineages, either similar or different in their phenotypical characteristics, and such cells can be selected by cloning or cell separation or by any other suitable procedure. The cell line obtained after selection or cloning is called a cell strain, which does not have a infinite life, since they die after a number of divisions.

1.8.1 Types of cell lines

As discussed above, cell lines that lose their ability to divide after a limited period of time are finite cell lines, i.e. these cell lines have a limited life span. Usually, finite cell lines contain cells which can divide 20–100 times (i.e. population doubling by 20–100 times) before losing their capability to divide. The extent of population doubling is dependent on several factors, such as cell lineage, cell type, origin, species, culture environment, etc. It has been noted that population doubling for human cell lines is between 50–100 times, whereas murine cell lines divide 20–30 times before extinction.

In an independent culture, continuous subculturing of cells or treatment of cells with carcinogens (chemicals), oncogenic viruses, etc, results in changes in phenotypical characteristics, in particular morphology, which can alter cells and lead to the development of cells that grow faster than normal cells. Cell lines obtained from these altered cells have infinite life spans. Such types of cell lines are referred to as continuous cell lines. These cell lines are immortal, transformed and tumorigenic (unlike the cell strains from which they were derived). Terms frequently used in animal tissue culture, in particular in the context of cell lines, are defined below:

- *Adherent cells.* Cells with the potential to adhere to the surface of the culture vessel using the extracellular matrix.
- *Immortalization.* Achieving a state of cell culture when cells proliferate continuously.
- *Attachment efficiency.* The proportion of cells that actually adhere to the surface of the culture vessel within a given time after inoculation.
- *Passaging.* The transfer of cells from one culture vessel to another. A more specific term is subculturing where the cells are first subdivided before being transferred into multiple cell culture vessels. A passage number will refer specifically to how many times a cell line has been subcultured. A number of adherent cell cultures will stop dividing when they become confluent (i.e. the stage when they entirely cover the surface of the cell culture vessel), and a number will die if they are retained in a confluent state for longer periods. Thus adherent cell cultures require repeated passaging, which means that when the cells are at the confluent stage, subculturing is required. Regular passaging is required in the case of suspension cultures, where suspended cells use their culture medium rapidly, particularly when the cell density becomes very high. While repeated passaging is essential to maintain cultures, the process is comparatively traumatic for adherent cells since they need to be trypsinized. Thus passaging of adherent cell cultures more than once every 48 h is not recommended.
- *Split ratio.* Divisor of the dilution ratio of a cell culture.
- *Generation number.* The number of doublings that a cell population has undergone. It should be observed that passage and generation number are not the same.
- *Population doubling time.* The population doubling (PD or pd) number is the estimated number of doublings that the cell population has undergone since isolation.
- *Passage number.* The number of times the culture has been subcultured.

1.8.2 Standard nomenclature of cell lines

The source and clone number (which represents the number of cell lines derived from the same donor) help in understanding the nomenclature more easily. The basic nomenclature is usually followed by assigning codes or designations to cell lines for their further identification, e.g. HeLa-S3 represents a human cervical tumor cell line, and similarly NHB 2-1 is a cell line derived from normal human brain (NB), followed by cell strain 2 and clone number 1. Another example is the MG-63 cell line. It is the 63rd sample of a tumor that produces a high amount of interferon beta. Therefore, its nomenclature is 'human tumor-63', or in Dutch, 'menselijk gezwell-63' or MG-63. Recently cell lines have transformed scientific study and are used for several purposes, such as:

- Vaccine production.
- Examining drug metabolism.
- Cytotoxicity.
- Antibody production.
- Investigating gene function.
- Development of artificial tissues (e.g. artificial skin).
- Production of biological compounds (e.g. therapeutic proteins).

Cell line requirements can be assessed through recent publications using specific cell lines. The American Type Culture Collection (ATCC) cell biology collection contains information on almost 3600 cell lines derived from 150 species. Although they are a useful tool, researchers must be careful when using cell lines instead of primary cells. The simultaneous use of cell lines and primary cells has been supported recently.

Cell lines should display and maintain functional features as close to the primary cells as possible. This may be particularly difficult to determine, as often the functions of the primary cells are not entirely understood. Since cell lines are genetically manipulated, this may alter their phenotype, native functions and responsiveness to stimuli. Serial passage of cell lines can further cause genotypic and phenotypic variation over an extended period of time, and genetic drift can also cause heterogeneity in cultures. Therefore, cell lines may not adequately represent primary cells and may provide different results. Additional problems include the chance of contamination with other cell lines and mycoplasma. In the early 1970s, cell line (inter- or intraspecies) mediated cross-contamination was explored by Nelson-Rees. Contamination of one cell line with a new one results in mixed cultures or occasionally complete overgrowth of the original cells by the contaminating line, and is an old problem. Nelson-Rees used chromosome banding (a procedure in which condensed chromosomes are stained to produce a visible karyotype) to prove that numerous immortal cell lines, earlier supposed to be unique, were in fact HeLa cell lines. He also demonstrated the fact that contamination with HeLa cells is responsible for the outgrowth of other cell lines [17–19]. Nelson-Rees demonstrated clearly that most of the of cell lines being investigated globally and distributed by cell banks [20] were contaminated with HeLa cells. This is the most considerable challenge for the animal tissue culture industry. During cell line contamination, contaminants, in particular rapidly proliferating cells, take over a whole cell line before its own growth takes place [21, 22]. HeLa cells are a frequent contaminant and, moreover, other contaminants such as mycoplasma can continue undetected in cell cultures for a long period of time. This prolonged exposure to contaminants can cause widespread changes in gene expression and cell behavior. According to certain reports, 15%–35% of cell lines submitted to cell banks were likely to be contaminated with mycoplasma [23, 24]. Thus appropriate precautions must be taken whenever cell lines are investigated.

1.8.3 Cell line selection

Usually the selection of high-producing cell lines is tedious and labor-intensive. High-producing cells are usually selected after transfection by using limiting dilution cloning to avoid non- and low-producing cells from outgrowing high-producing cells. This process usually takes more than three months. During this time, the cells have to be screened occasionally to ensure stability of the selected clone. High-producing mammalian cell line selection is one of the considerable challenges in the production of biopharmaceuticals. Increasing demand for therapeutic proteins requires the urgent development of methods for the selection of mammalian cell lines stably expressing recombinant products at high levels in an efficient, cost-effective and high-throughput manner [25–27]. Numerous approaches for selecting and screening cells have been explored, including flow cytometry, gel microdrop methods (encapsulating the cells in gelatin beads) and matrix based secretion assays. Recently, fluorescence-activated cell sorting has been utilized to estimate the cell-specific productivity (Qp), or the quantity of product produced per cell per day [25–27]. This parameter is utilized in biopharmaceutical cell selection in a cell-specific manner, which allows multi-purpose characterization and isolation of individual cell clones from heterogeneous populations. Several factors are considered during the selection of cell lines, such as [25–27]:

- Origin of the cell line (human or non-human cell line; human cell lines are more vulnerable to different types of contamination).
- Type of cell line (finite or continuous).
- Types of cells (normal or transformed).
- Growth patterns.
- Cloning efficiency.
- Cell number under specified culture conditions (saturation density).
- Population doubling time.
- Availability of cell line.
- Availability of growth factors or media for its maintenance.
- Physical expression of traits, or characteristics.

1.8.3.1 Quarantine

To avoid microbial contamination, new cell lines should be quarantined (kept entirely separate from existing cell line stocks). Usually, an independent quarantine laboratory should established for this purpose. A class-II microbiological safety cabinet (MSC) and an incubator dedicated to quarantine can be considered as an alternative approach.

1.8.4 Verification of a cell line

To confirm the origin of a cell line and to avoid misidentification, cell line authentication is carried out using an established DNA based method. With the advancements in tissue culture science, it is now possible to equate the cell line DNA with that of the tissue of origin, however, this is only possible for a few cell lines that are already available. Short tandem repeat analysis can be used to confirm the origin of a cell line. A short tandem repeat pattern is derived and compared for the cell line and the primary culture. The primary culture should be frozen or processed so that it can clearly be determined that the cell line is obtained from a recognized donor. This method is recommended for purposes of authentication, so that the unique identity of the primary culture is available for the international database (NCBI 2013) [28]. Some other methods, such as genotypic methods (karyotype, copy number variation mapping or even whole-genome sequencing) can also be used to authenticate cell lines.

1.8.5 Characterization of cell lines

Before characterization, the handler first ensures that the cell line obtained is appropriate or acceptable for their designed experiment or purpose. Even after confirmation of the cell line, it is essential to check whether the cell line is still carrying key characteristics after persistent passaging. To reveal changes in a cell line, karyotyping is the most recommended approach. It can demonstrate that a cell line has a normal karyotype, and the cell line can then be used for various research purposes.

Karyotyping is thus a simple test that can reveal changes in a cell line. Indeed, it is routine to demonstrate that a line of embryonic stem cells or induced pluripotent stem cells has a normal karyotype if they are to be used for experiments involving the production of chimeras and germ line transmission. Molecular assays for copy number variation or RNA profiling will also be indicative of changes, but are more costly. Nevertheless, a great deal of time and effort can be saved by confirming the presence of appropriate characteristics before commencing work. It is also advisable to capture an image of the cell line in culture at different cell population densities and perform basic characterization (e.g. calculating the population doubling time for that cell line) soon after arrival. For a newly developed cell line it is imperative to authenticate the origin of the cell line and the extent of variation between cells present in the primary tissue culture.

1.8.6 Misidentification of cell lines

Cross-contamination is considered as the primary cause of misidentification. A high risk of cross-contamination is usually associated with continuous cell lines as they may replace other, slow growing cell lines. The following lists a number of factors that are responsible for misidentification:

- Alterations in cellular behavior or morphological variations.
- Developing two cell lines in an Mesenchymal stem cells simultaneously.
- Failure in maintaining good cell culture practice.
- Liquefying the wrong ampoule.
- Mislabeling a flask or ampoule.

- Persistence of the mitotic activity of feeder cells such as embryonic stem cells because of insufficient irradiation or treatment with mitomycin C.
- Poorly controlled manipulation.
- Unintended transfer of cells to a stock bottle of medium.
- Using unsterilized media (used without suitable filtration to remove cells).

1.8.7 Maintenance of a cell line

Current research practice demands the development of good models, as good science cannot be achieved with bad models. Several cell culture procedures have been developed in the current century which overcome the drawbacks of traditional culture procedures and are more scientifically rigorous, such as stem cell derived human cells, co-cultures of different cell types, scaffolds and extracellular matrices, tissue architecture, perfusion platforms, organ-on-chip technologies, 3D culture and organ functionality. The biological relationships between such models can be further improved by organ-specific approaches, more widespread assessment of cell responses using high-content methods and by using biomarker compounds. These strategies can be utilized to make a microphysiological model system. One of the most significant advantages of this type of model system is that it generates results closer to the *in vivo* situation, however, controlling multiple parameters is considered a significant challenge for animal tissue culture industries. Cell line maintenance has become a very valuable undertaking, both in academic research and in industrial biotechnology. The following factors should be considered during maintenance of a cell line in culture.

1.8.7.1 Cellular morphological examination

Cells should be examined routinely to check for the presence of any other contaminant. Morphological examination is essential to investigate and differentiate the natural cellular organization and the physiological state of the cells from the contaminated. Therefore, morphological examination is usually used as a qualitative and quantitative measure of various biological assays.

1.8.7.2 Media replacement

Regular changing of the medium is required to maintain cell lines in culture; however, the frequency of changing the medium always varies. For example, proliferating cells require more nutrients in comparison to non-proliferating cells. The rate of cellular growth and metabolism decides the interval between the changing, or addition of fresh, medium. To understand this better, we can consider HeLa, rapidly growing transformed cells. In order to avoid contamination and to meet cell nutrient requirements, HeLa cell medium should be replaced twice in a week, whereas for slow growing cells (non-transformed cells), e.g. IMR-90, the medium can be replaced once a week. Thus, rapidly growing or proliferating cells require more frequent changes of medium than slow growing or non-proliferating cells. Several factors should be considered when changing the medium:

- *Cell density.* Cultures with a high density of cells use medium faster than those with low density, thus medium need to be changed more frequently for high densites of cells.
- *Fluctuations in pH.* Changes in pH should be monitored carefully, since a decrease in pH may be associated with a decline in the growth rate of the cells. The optimal pH for the growth the cells is 7, and a decline in pH (6.5) can retard the growth of cells. Further decline in pH (usually to between 6.0–6.5), may stop the growth of the cells, and if the low pH persists cells will start losing their viability. Thus the pH should be carefully monitored for each cell line and controlled with a suitable medium. Medium changing is not required when the pH declines by 0.1 units/day, as such a decline may not harm cells, however, declines of 0.4 pH units/day may affect growth, and eventually the viability of cells, so in this case an immediate medium change is required.
- *Type of cell.* Feeder cells such as embryonic stem cells and tumorigenic cells such as transformed cell lines (continuous cell lines) grow fast and thus require a greater supply of nutrients. Thus rapidly growing cells require more frequent medium changes than normal cells.
- *Phenotypical variations.* It is important to examine cell morphology carefully using specific techniques, as any change in morphology could be a sign of contamination or deterioration which can ultimately affect the growth of the cells.

1.9 Subculture

Subculture is defined as the transfer of cells from one culture to start another culture. During this process proliferating cells are subdivided, which allows the development of new cell lines. This step is referred to as a passage, and a passage number is the recorded number of times a cell culture has been subcultured. Numerous adherent cell cultures will stop proliferating when they reach the confluent stage (i.e. when they completely cover the surface of the cell culture vessel), and certainly will die if they are left in the confluent stage for a prolonged period. Thus adherent cell cultures should be regularly passaged, that is, when cells reach the confluent stage a portion of the cells need to be passaged or subcultured to a new cell culture vessel. However, it is not recommended to regularly subculture adherent cells (no more than once every 48 h) as they must be trypsinized. In contrast, suspension cultures with high cell density require routine passaging as they use medium rapidly.

The standard growth curve of cells in a culture is shown in figure 1.2.

During the initial lag phase there is less growth as the cells are not adapted to the environment. Once they start adapting to the environment they proliferate exponentially, which is why this is called the exponential or log phase. This is the time when all cells actively grow and consume medium. During this time the medium should be changed, otherwise growth will stop. As discussed above, the confluent phase is reached when the culture exceeds the capacity of the medium. At this stage the culture has to be divided into subcultures. There are two types of subcultures: monolayer and suspension subcultures.

1.9.1 Monolayer cultures

Monolayer cultures involve anchorage-dependent cells which can be established from human tissue after enzymatic treatment to disperse them into single cells in order to form a monolayer or single-cell continuous layer over the bottom of the vessel. Cellular attachment between cells and with the interface is facilitated by surface glycoproteins and calcium ions. Several quantitative approaches have been explored for examining viable cells in monolayer cultures, such as:

- Microscopic screening to examine morphological changes.
- Cytotoxicity studies.
- Incubation with dye followed by colorimetric analysis.

The initial step in subculturing of monolayers is to remove cells from the interface of the vessel by trypsinization or mechanical means [30]. The final dispersion is then subdivided and transferred to fresh cultures. The growth of the secondary cultures is periodically monitored and further subcultured to produce tertiary cultures, etc. As discussed above, the time interval between subculturings is entirely dependent on the growth rate and varies with the cell line.

1.9.2 Procedures for cell detachment

There are various means of cellular detachment from the culture vessel interface, such as physical and chemical methods (figure 1.3).

The utilization of proteases is not recommended when cultures are loosely adhered, thus mechanical shaking and scraping are more appropriate in such cases. Due to its advantages, trypsin is often used for cell dissociation, however, other enzymes such as pronase, dispase and collagenase are used when monolayers cannot be disaggregated with trypsin. Prior treatment with EDTA is required to remove Ca^{+2} so that it will not interfere with the action of enzymatic dissociation, and eventually uniform dispersion can be achieved [30]. As one-cell-thick monolayers are the simplest tissues in multicellular organisms, they act as a suitable model for

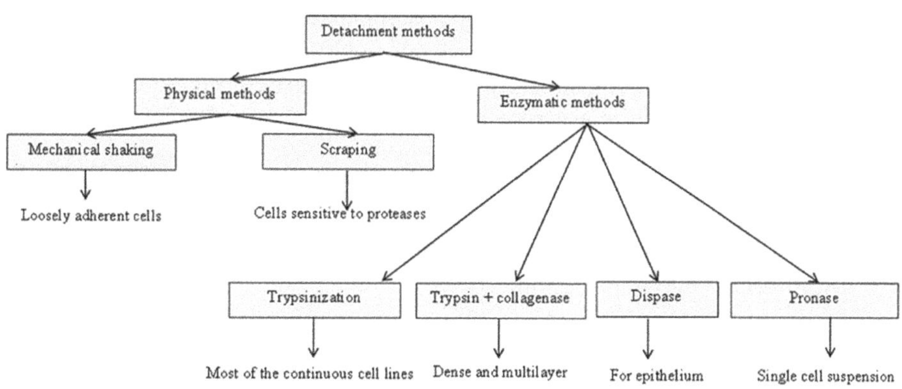

Figure 1.3. Methods for dissociation of cells.

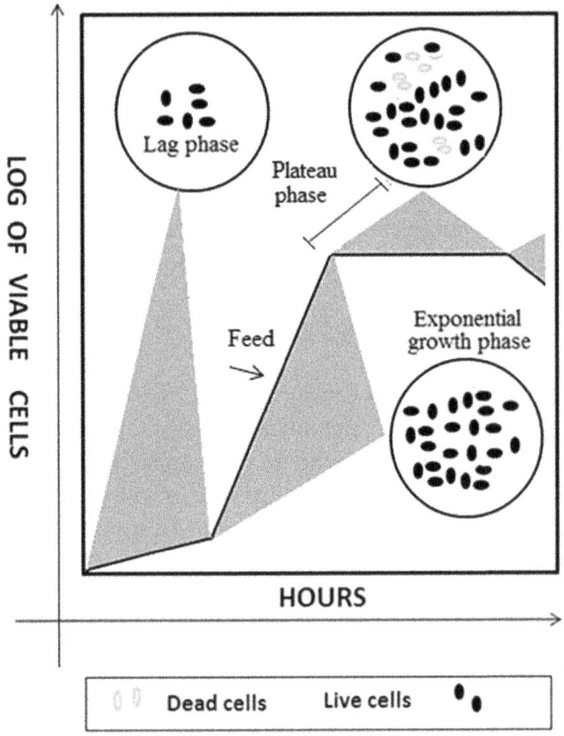

Figure 1.4. Subculturing.

development and normal physiology. It has been determined that the extracellular material (ECM) should be carefully considered when selecting the dissociation approach, as this helps in determining the effects of the dissociating agent on the cytoskeleton, adherent junctions and desmosomes. Usually monolayers can withstand the different mechanical stresses exerted by the interface itself under *in vitro* conditions, and can shield the internal environment from harmful external elements. Since dissociating elements or external environmental factors can affect ECM synthesis, optimization of the dissociating element is required before treatment in order to estimate a suitable dose for dissociation [30].

As mentioned in figure 1.4, subculturing is usually carried out between the middle log and plateau phases; it is not recommended to start subculturing during the lag phase.

Understanding of growth patterns is necessary for:

- Designing culture experiments.
- Regular maintenance of a culture.
- Monitoring cell proliferation.
- Evaluating a culture's response to external factors.

There are certain considerations in subculturing monolayers, as follows.

Cellular density

Subculturing time is dependent on the cellular density. Cellular density is usually found to be high at the confluence stage. So, whenever normal or transformed cells reach the confluence stage it is advisable to perform subculturing as this can maintain the balance between nutrient supplementation and consumption by the cells/micro-organisms. In the confluence stage, when all the growth area is utilized and cells start coming closer to each other, growth may be hampered due to the negative force (contact inhibition) developed between the cells striving for nutrients to further meet their energy demands.

Exhaustion of nutrients

Usually, in microbiology a sudden drop in pH represents an increase in cellular density which again signifies the confluence stage, thus a drop in pH often necessitates subculturing.

Reason for subculture

Subculturing is also done in those cases when cells are to be used for any specific purpose other than routine propagation, in order to obtain high yield or stock, or to change the type of medium. In such cases the cell has to be subcultured frequently.

Scheduled timings for subculture

As we know, regular subculturing is generally performed as per a strict schedule to obtain significant results. Seeding density should be increased in the case that the cell will not reach the confluence stage at a suitable time, and seeding density should be reduced when the cells will reach the confluence stage a little too early. It is now possible to determine the correct seeding density and subculture interval by studying standard growth curves. In most cases medium change is performed after 3–4 days and subculturing after 7 days.

The steps involved in monolayer subculture are shown in figure 1.5. Monolayer subculturing involves multiple steps:

- Medium removed and monolayer washed.
- Treatment of cell with trypsin.
- Trypsin removed leaving a residual film.
- Incubation (37 °C for 30 min).
- Cell rounding up after incubation.
- Resuspension of the cells in the medium.
- Reseeding of cells.
- Confluent stage of monolayer.

For the majority of continuous cell lines, the seeding concentration for subculturing lies between 1×10^4 to 5×10^4 ml. However, for developing a new culture, the initial concentration should be high and should then be reduced to meet the culture's requirements.

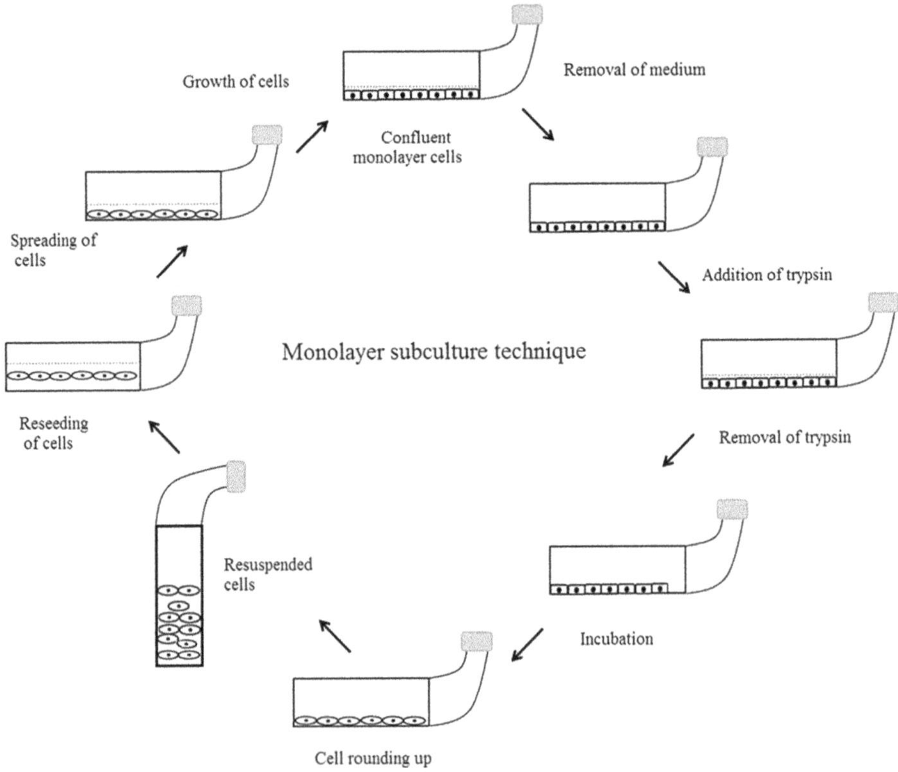

Figure 1.5. Process of monolayer subculture.

1.10 Suspension cultures

Most cell lines are grown as monolayer adherent cells, which grow only on the surfaces of culture vessels, however, certain cells are not adhesive, such as cells derived from leukemic tissue. Moreover, certain cells do not require support for their growth. These cells can be mechanically kept in suspension, and such cultures are referred to as suspension cultures. Transformed cells are usually subcultured using this method. Suspension culture of animal tissues is similar to the method used to subculture bacteria or yeast. There are a number of advantages to suspension cultures over monolayer cultures:

- Bulk production or production in mass can be achieved.
- The cultured cell has access to nutrition from all directions.
- Easy to maintain.
- Frequent replacement of medium is not required.
- The lag period is short.
- The process of propagation is fast.
- Scale-up is convenient.
- Trypsin treatment, or any other enzyme treatment, is not required.

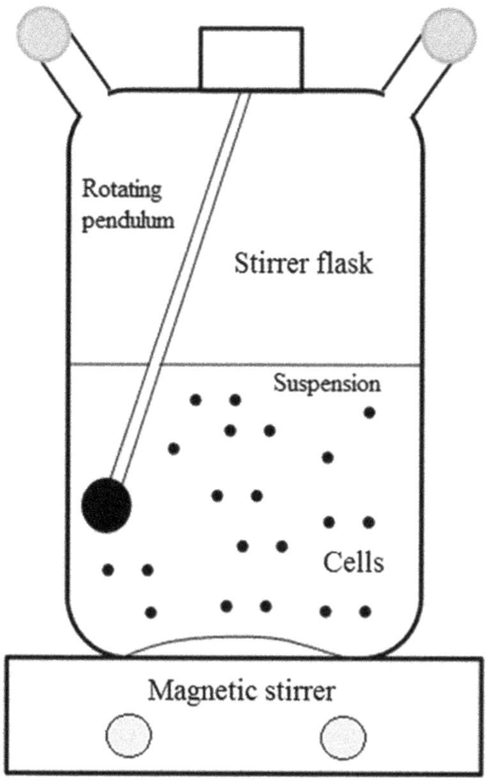

Figure 1.6. Stirred flask technique for suspension cultures at a large scale.

Similar parameters have been reported for suspension cultures as for monoculture, i.e. culture density, fluctuations in pH, schedule timings, the purpose of the subculture, etc. During this process cells are suspended in a culture flask (figure 1.6) which contains culture medium. In the stirred flask technique, the medium is continuously stirred with the help of a magnetic pendulum, in order to offer homogeneous stirring and to avoid aggregation. This magnetic pendulum is allowed to rotate at the base of the flask and the suspension of cells should be regularly monitored for contamination, aggregate formation or any signs of deterioration.

1.10.1 Cell synchronization

Synchronized cells have the same growth rate in all generations, whereas unsynchronized growth means different growth rates of cells, as shown in figure 1.7. The cell culture has to be synchronized so that the cells will be at the same phase at the same time, which makes it easier to determine the growth rate. Cell synchrony is essential to study the development of cells through the cell cycle, which needs to be monitored at periodic intervals. Several methods have been introduced to achieve cell synchronization. These approaches are broadly divided into two categories:

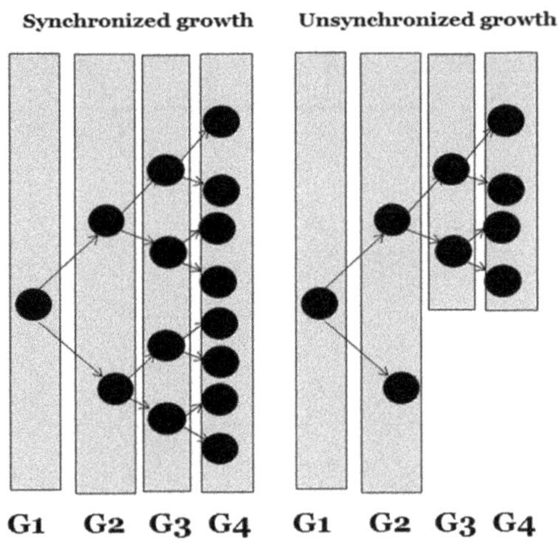

Figure 1.7. Simple illustration of synchronized (cells are dividing at the same time) and unsynchronized growth (cells are not dividing at the same time).

- Cell synchronization by physical means.
- Cell synchronization by chemical means.

Cell synchronization using physical methods is more effective than using chemical methods, as the latter can cause toxicity to the cells. Deprivation of nutritional resources (part of the chemical approach) cannot be utilized to synchronize transformed cells. As the cell cycle is composed of different development or growth phases, these determine the synchrony of the cells, and more synchrony can be obtained at the first cycle than at the second or third cycles.

1.10.2 Cell synchronization by chemical means

In this approach cells are synchronized by blocking metabolic reactions, which can be achieved by either adding inhibitor substances to the culture medium or by depriving the micro-organisms or cells of nutritional sources.

Inhibitors such as thymidine, aminopterine, hydroxyurea, cytosine and arabinoside, which have variable effects, are utilized to block DNA synthesis during the S phase of the cell cycle, bringing the cells to the same phase.

Removing essential growth substances, such as serum or isoleucine, from the culture medium for almost 24 h leads to the accumulation of cells at the G1 phase. This approach of depriving cells of certain nutritional components exposes the cells to similar types of stress in response to which the cells will present similar adaptations, and thus synchrony can be achieved.

1.10.3 Cell synchronization by physical means

Separation of cells by physical means to achieve synchrony can be done using characteristics such as cell density, affinity against antibodies, light scattering or fluorescent emission by labeled cells. Techniques which can be commonly utilized to separate cells based on their adaptations and phenotypical variations are centrifugal elutriation and fluorescence-activated cell separation.

Centrifugal elutriation is a process to enhance the sedimentation rate to improve the yield of cells. The process is based on cell size and sedimentation velocity. During this process cells in the medium are ejected into the separating chamber such that they will be forced to the edges. This occurs in such a way that the centripetal force will be equivalent to the sedimentation rate of the cells. As the cell present in the culture must have phenotypical variations, such as size, shape, density, cell surface, etc, thus cells at different phases of the cell cycle have a tendency to sediment at different rates and different positions in the chamber. The whole process can be monitored via a porthole, as the chamber is illuminated by stroboscopic light.

1.11 Algal extracts in animal tissue culture

Algal extracts contain high amounts of potential secondary metabolites which can be utilized to either elicit or promote growth of cells under *in vitro* conditions. These metabolites can be utilized in media to further increase the growth of the cells. Our recent research on the red algae, *Porphyra vietnamensis*, found among its diverse chemical compounds some with significant pharmacological properties [31–50]. Such types of algae can be utilized to elicit the growth of animal cells. Shinohara *et al* observed that algal phycocyanins are responsible for the growth of human cells in culture [50]. In their study, growth-promoting substances derived from blue-green algae, *Synechococcus elongatus* var., were separated to produce a biliprotein fraction that promoted the growth of RPMI 8226 cells; allophycocyanin was found to be more active than phycocyanin.

1.12 Animal and plant tissue culture

Tissue culture is the art of growing cells outside a living body. As we have already discussed the historical background and current innovations of the field of biotechnology in the first two volumes, it is well understood that there are direct and indirect relationships between the developmental biology of plant and animal cells [47]. There are certain challenges involved in culturing both plant and animal cells, such as exhaustion of nutrients in the growth medium, apoptotic/necrotic cell accumulation, cell cycle arrest (or senescence) due to intercellular communication or contact inhibition, etc, which should be investigated further [47]. Various approaches can be utilized to manipulate cell cultures of both plant and animal cells. Subculturing is a common practice which is adopted to replace old medium with new, nutrient enriched, medium. Subculture can also be used to prevent the major problem of senescence. This involves transferring a small number of cells into a new culture dish. The animal–plant co-culture system has not been explored due to its greater vulnerability to contamination of the culture. There is, however, scope to maintain suitable aseptic conditions and encourage a co-culture system to further study the impacts of their growth on each other [47].

The success of animal tissue culture products depends on their efficacy, cost effectiveness and the potential for scale-up. Recent and current advances in tissue culture science have enhanced the complexity in the design of biomaterials which have either been proposed or utilized to grow animal cells [51]. This complexity generally increases the difficulty for manufacturing industries in designing suitable fabrication techniques. It is worth noting that most of the features that are suitable for designing biomaterials originate in the structure and function of plants [51]. Several investigations have shown that decellularized plant tissues can be utilized as a suitable scaffold for the culture of human cells. It has been observed that through an approach of simple biofunctionalization it is possible to achieve the adhesion of human cells on various sets of plant tissues. The increased water transport efficiency and hydrophilicity of plant tissues facilitate increases in cell number over prolonged periods of culture [51]. In addition, animal cells are able to adapt well to the microstructure of plant frameworks without breaking any physiological conditions. This results in perfect cell positioning and formation of a perfect pattern over the feeding layer of plant cells. This supportive plant tissue based micro-framework can be utilized as an alternative potential scaffold for mammalian cells [51].

1.13 Biomaterials and animal tissue culture

Recently, there has been a boom in biopolymers of natural and synthetic origin [51–53]. It is essential to understand the possible interactions between cells and these materials in order to develop new materials [48, 54]. When isolated cells from suitable tissues are cultured on a plastic culture dish, this transition of cells from an *in vivo* to *in vitro* environment results in the loss of several functions and cells usually begin dedifferentiation, for unknown reasons. Identification of the microenvironmental signals that are responsible for changes in cellular phenotype and function will help in understanding cell behavior under *in vitro* conditions. Most of the current research that deals with tissue-engineered constructs involves suitable scaffolds that not only act as anchorage cells, but that also help in studying cell behavior and developmental stages in more detail [48, 54]. Several scaffolds have been developed that offer suitable architecture, in particular initial structural integrity and support or a backbone in the form of a matrix in which the cells arrange themselves and form a mass of functioning tissue [51–53]. Several techniques have also been developed, such as three-dimensional matrices for culturing animal cells, to help in understanding how cells probe their surroundings [51–53]. The biomaterial matrix is designed in such a way that it will be able to control the cell position and function inside artificial environments [48, 54]. For most of the materials developed so far, we lack understanding of the effects of the biomaterial or surrounding microenvironment on cell development, behavior and functions.

Physiologically, cells are always surrounded by a sophisticated and dynamic microenvironment which includes the extracellular matrix, growth factors and cytokines, as well as neighboring cells. The extracellular matrix helps in connecting the cells' extracellular matrix proteins through specific cell surface receptors such as integrins [55–57]. Such receptors are responsible for connecting the intracellular cytoskeleton to the extracellular matrix [48, 54]. It is important to understand the interaction between the ligands present in the extracellular matrix and the receptors of the cell. Such an interaction allows multiple intracellular signaling processes that can result in the alteration of cellular behaviors, such as growth, migration and differentiation. Natural extracellular matrices such as collagen offer natural adhesive ligands that encourage cellular connection with integrins. Such biomaterials can be considered as potential resources for engineering new biomaterials [55–57]. One of the major shortcomings of such natural extracellular matrices is our failure to control their physicochemical properties. Several natural biomaterials have been explored recently. In one of the recent innovations in ligand chemistry, a short peptide sequence (arginine–glycine–aspartic acid) which is responsible for cellular adhesion was discovered [55–57]. This peptide can be conjugated with other biologically inert polymers to study their effect on cultured animal cells [55–57]. These peptides, once conjugated over the matrix of inert biomaterial, can initiate cellular adhesion which can further allow researchers to develop suitable matrices on the surface of which these peptides can be conjugated. This approach allows the development of suitable matrices the adhesion potential and chemistry of which can be controlled. Additionally, the incorporation of growth factors in the matrix facilitates their distribution to cells in a controlled fashion. Therefore, cellular adhesion and growth of the cell can be controlled by changing the chemical nature of the polymeric network. One of the most common approaches is functionalization. Growth factors can also be immoblized over polymeric networks to study and manipulate cells. One of the common examples of such immobilization is used in the study of insulin and epidermal growth factor [55–57].

Pattern-immobilization is a more reliable approach as in this process cells are allowed to culture in a matrix with an architecture that produces a three-dimensional structure mimicking the *in vivo* environment [55–57]. In such a systematic spatial arrangement, growth factor proteins or other proteins, such as those carrying the ligands, are embedded to encourage interaction between cell receptors and ligands. Such artificial frameworks help in the development of tissue through non-diffusion mechanisms, in which the movement of proteins is not dependent on the concentration gradient. This type of stimulation by immobilized growth factors mimics the *in vivo* environment of membrane-anchored growth factors such as heparin-binding epidermal growth factor, transforming growth factor and tumor necrosis factor. Furthermore, cellular growth can also be enhanced by co-immobilization with adhesion factors and by means of thermosensitive polymers, which allows the development of cells and, most importantly, allows the recovery of the cell through reducing the temperature [55–57].

1.14 Nanotechnology and biotechnology

Three dimensional biomaterials with large pore size (greater than 100 μm) carry a high number of functional units essential for the regeneration of various tissues. Pore size greater than 100 μm is essential for the cell adhesion and proliferation, whereas biomaterials with 325 μm pore size encourage migration of cells through the scaffolds. Scaffolds with pore size less 85 μm showed lowest intensity of cell adhesion and migration. So far, most tissue engineering studies have focused on macro-sized frameworks for cells greater in size than 100 μm (subcellular size) or cellular arrangements larger than 10 μm (cellular size). Such massive structures are required to produce real-sized organ systems [49]. However, to design functional units of tissue, not only are the subcellular and cellular scales required, but also nanostructures, 1–100 nm in size. This type of structural arrangement is essential to control cell behavior, in particular cell–cell interactions, cell–molecular interactions and the cellular environment [49]. The recovery of cell characteristics, in particular structure and function, can only be achieved by reconstruction of the nanostructures of the tissue itself. The current prospects of tissue engineering are very dependent on understanding the interaction of cells with these nanostructures. These tiny structures with three-dimensional arrangements can directly or indirectly affect the cell functions [49]. The trend to fabricate ever smaller structures (called miniaturization), mainly to regenerate the components inside the targeted tissue, has proven a reliable approach for researchers. Thus far, several nano-materials have been developed to mimic native tissues. These nanostructures are tissue-engineered grafts, biomaterial scaffolds that are engineered and then manufactured at the molecular level [49]. Several techniques are available to optimize materials, even at the level of atoms, molecules and supermolecules, 1–100 nm in scale. By means of nanotechnology, various materials or devices can be fabricated or designed to offer a product with high biocompatibility, and most importantly highly predictable biological and physical properties [49]. Animal biotechnology is a broad discipline that includes DNA science, genetic engineering, transgenic science and stem cell research. DNA research involves DNA isolation and screening methods, whereas genetic engineering involves the manipulation of the genetic makeup of an organism to either synthesize a product or alter the character of the organism [58, 59]. Currently, all these fields are being utilized to derive biopharmaceuticals. Moreover, enzyme research, mainly protein engineering, immobilization and biotransformation, has several applications in animal tissue culture science [58, 59].

References

[1] Rodríguez-Hernández C O, Torres-Garcia S E, Olvera-Sandoval C, Ramirez-Castillo F Y, Muro A L and Avelar-Gonzalez F J 2014 Cell culture: history, development and prospects *Int. J. Curr. Res. Acad. Rev.* **2** 188–200

[2] del Carpio A 2014 The good, the bad, and the HeLa *Berkley Sci. Rev.* 5

[3] Masters J R 2002 HeLa cells 50 years on: the good, the bad and the ugly *Nat. Rev. Cancer* **2** 315–9

[4] Schwarz E, Freese U K, Gissmann L, Mayer W, Roggenbuck B, Stremlau A and zur Hausen H 1985 Structure and transcription of human papillomavirus sequences in cervical carcinoma cells *Nature* **314** 111–4

[5] Wezel A L, Steenis G, Hannik Ch A and Cohen H 1978 New approach to the production of concentrated and purified inactivated polio and rabies tissue culture vaccines *Dev. Biol. Stand.* **41** 159–68

[6] Stones P B 1977 Production and control of live oral poliovirus vaccine in WI-38 human diploid cells *Dev. Biol. Stand.* **37** 251–3
[7] Beale A J 1981 Cell substrate for killed poliovaccine production *Dev. Biol. Stand.* **47** 19–23
[8] Steenis G, Wezel A L, Groot I G M and Kruijt B C 1980 Use of captive-bred monkeys for vaccine production *Dev. Biol. Stand.* **45** 99–105
[9] Shenoy M 2007 Animal cell culture *Animal Biotechnology* (New Delhi: Firewall) ch 1, p 3
[10] Hayflick L and Moorhead P S 1961 The serial cultivation of human diploid cell strains *Exp. Cell Res.* **25** 585–621
[11] Wiktor T J, Fernandes M V and Koprowski H 1964 Cultivation of rabies virus in human diploid cell strain WI-38 *J. Immunol.* **93** 353–66
[12] Capstick P B, Telling R C, Chapman W G and Stewart D L 1962 Growth of a cloned strain of hamster kidney cells in suspended cultures and their susceptibility to the virus of foot and mouth disease *Nature* **195** 1163–4
[13] Losee J R and Ebeling A H 1914 The cultivation of human sarcomatous tissue *in vitro J. Exp. Med.* **20** 140–8
[14] Earle W 1943 Production of malignancy *in vitro*. IV. The mouse fibroblast cultures and changes seen in the living cells *J. Natl Cancer Inst.* **4** 165–212
[15] Pullen K, Johnston M D, Philips A W, Ball G D and Finter W B 1984 Very large scale suspension cultures of mammalian cells *Dev. Biol. Stand.* **60** 175–7
[16] Geraghty R J *et al* 2014 Guidelines for the use of cell lines in biomedical research *Br. J. Cancer.* **111** 1021–46
[17] Gómez-Lechón M J, Donato M T, Castell J V and Jover R 2003 Human hepatocytes as a tool for studying toxicity and drug metabolism *Curr. Drug Metab.* **4** 292–312
[18] MacDonald C 1990 Development of new cell lines for animal cell biotechnology *Crit. Rev. Biotechnol.* **10** 155–78
[19] Schurr M J *et al* 2009 Phase I/II clinical evaluation of StrataGraft: a consistent, pathogen-free human skin substitute *J. Trauma* **66** 866–73
[20] Nelson-Rees W A, Daniels D W and Flandermeyer R R 1981 Cross-contamination of cells in culture *Science* **212** 446–52
[21] Capes-Davis A *et al* 2010 Check your cultures! A list of cross-contaminated or misidentified cell lines *Int. J. Cancer* **127** 1–8
[22] Jiang L, Zeng X, Wang Z and Chen Q 2009 Cell line cross-contamination: KB is not an oral squamous cell carcinoma cell line *Eur. J. Oral. Sci.* **117** 90–1
[23] Fleckenstein E, Uphoff C C and Drexler H G 1994 Effective treatment of mycoplasma contamination in cell lines with enrofloxacin (Baytril) *Leukemia* **8** 1424–34
[24] Hay R J, Macy M L and Chen T R 1989 Mycoplasma infection of cultured cells *Nature* **339** 487–8
[25] Carroll S and Al-Rubeai M 2004 The selection of high-producing cell lines using flow cytometry and cell sorting *Expert Opin. Biol. Ther.* **4** 1821–9
[26] Browne S M and Al-Rubeai M 2007 Selection methods for high-producing mammalian cell lines *Trends Biotechnol.* **25** 425–32
[27] Gallagher C and Kelly P S 2017 Selection of high-producing clones using FACS for CHO cell line development *Methods Mol. Biol.* **1603** 143–52
[28] NCBI Resource Coordinators 2015 Database resources of the National Center for Biotechnology Information *Nucleic Acids Res.* **44** D7–19
[29] Borenfreund E and Puerner J A 1985 Toxicity determined *in vitro* by morphological alternation and neutral absorption *Toxicol. Lett.* **24** 119–24

[30] Harris A R, Peter L, Bellis J, Baum B, Kabla A J and Charras G T 2012 Mechanics of cultured cell monolayers *Proc. Natl Acad. Sci.* **109** 16449–54
[31] Bhatia S, Goli D, Naved T and Sharma A 2018 Nutraceutical properties of Indian seaweed porphyra *Adv. Invest. Pharmacol. Ther. Med.* **1** 47–54
[32] Bhatia S 2016 Nanoparticles, plants, and algae *Natural Polymer Drug Delivery Systems* (Cham: Springer) pp 117–27
[33] Bhatia S and Goli D 2016 *Leishmaniasis: Biology, Control and New Approaches for Its Treatment* (Boca Raton, FL: CRC Press) pp 164–73
[34] Bhatia S, Namdeo A G and Nanda S 2010 Factors effecting the gelling and emulsifying properties of a natural polymer *Sys. Rev. Pharm.* **1** 86–92
[35] Bhatia S 2018 *Introduction to Pharmaceutical Biotechnology: Enzymes, proteins and bioinformatics* vol 2 (Bristol: IOP Publishing) pp 172–7
[36] Bhatia S, Garg A, Sharma K, Kumar S, Sharma A and Purohit A P 2011 Mycosporine and mycosporine-like amino acids: a paramount tool against ultra violet irradiation *Pharmacogn. Rev.* **5** 138–46
[37] Bhatia S, Rathee P, Sharma K, Chaugule B B, Kar N and Bera T 2013 Immuno-modulation effect of sulphated polysaccharide (porphyran) from *Porphyra vietnamensis Int. J. Biol. Macromol.* **57** 50–6
[38] Bhatia S *et al* 2008 Novel algal polysaccharides from marine source: porphyran *Pharmacogn. Rev.* **2** 271–6
[39] Bhatia S, Sharma K, Namdeo A G, Chaugule B B, Kavale M and Nanda S 2010 Broad-spectrum sun-protective action of Porphyra-334 derived from *Porphyra vietnamensis Pharmacogn. Res.* **2** 45–9
[40] Bhatia S 2018 *Introduction to Pharmaceutical Biotechnology: Basic techniques and concepts* vol 1 (Bristol: IOP Publishing) pp 167–74
[41] Bhatia S *et al* 2014 Significance of algal polymer in designing amphotericin B nanoparticles *Sci. World J.* **2014** 564573
[42] Bhatia S *et al* 2015 Investigation of the factors influencing the molecular weight of porphyran and its associated antifungal activity *Bioact. Carbohydr. Dietary Fibre* **5** 153–68
[43] Bhatia S, Sharma K, Sharma A, Nagpal K and Bera T 2015 Anti-inflammatory, analgesic and antiulcer properties of *Porphyra vietnamensis Avicenna J. Phytomed.* **5** 69–77
[44] Bhatia S, Sharma K and Bera T 2015 Structural characterization and pharmaceutical properties of porphyran *Asian J. Pharm.* **9** 93–101
[45] Bhatia S and Bera T 2015 Evaluation of pharmacognostical, phytochemical and anti-microbial properties of *Porphyra vietnamensis Int. J. Green Pharm.* **9** 131–7
[46] Bhatia S, Sharma K, Sharma A, Namdeo A G and Chaugule B B 2011 Anti-oxidant potential of Indian porphyra *Pharmacology* **1** 248–57
[47] Bhatia S *et al* 2015 *Modern Applications of Plant Biotechnology in Pharmaceutical Sciences* (New York: Academic, Elsevier) pp 164–74
[48] Bhatia S 2016 *Systems for Drug Delivery* (Berlin: Springer) pp 122–7
[49] Bhatia S 2016 *Nanotechnology in Drug Delivery: Fundamentals, Design, and Applications* (Boca Raton, FL: CRC Press)
[50] Shinohara K *et al* 1988 Algal phycocyanins promote growth of human cells in culture *In Vitro Cell. Dev. Biol.* **24** 1057–60
[51] Khan F and Tanaka M 2017 Designing smart biomaterials for tissue engineering *Int. J. Mol. Sci.* **19** 17

[52] Lee E J, Kasper F K and Mikos A G 2014 Biomaterials for tissue engineering *Ann. Biomed. Eng.* **42** 323–37

[53] Knight E and Przyborski S 2014 Advances in 3D cell culture technologies enabling tissue-like structures to be created *in vitro J. Anat.* **227** 746–56

[54] Bhatia S 2016 *Natural Polymer Drug Delivery Systems* (Cham: Springer)

[55] Fontana G *et al* 2016 Biofunctionalized plants as diverse biomaterials for human cell culture *Adv. Healthc. Mater.* **6** 1601225

[56] Liua W F and Chen C S 2005 Engineering biomaterials to control cell function *Mater. Today* **8** 28–35

[57] Ito Y 2002 Cell culture engineering using intelligent biomaterials *Animal Cell Technology: Challenges for the 21st Century* ed K Ikura *et al* (Dordrecht: Springer)

[58] Bhatia S 2018 *Introduction to Pharmaceutical Biotechnology: Basic Techniques and Concepts* vol 1 (Bristol: IOP Publishing)

[59] Bhatia S 2018 History, scope and development of biotechnology *Introduction to Pharmaceutical Biotechnology* vol 2 (Bristol: IOP Publishing)

IOP Publishing

Introduction to Pharmaceutical Biotechnology, Volume 3
Animal tissue culture and biopharmaceuticals
Saurabh Bhatia, Tanveer Naved and Satish Sardana

Chapter 2

Organ culture

2.1 Organ culture

The technique of *in vitro* culture which involves the maintenance or growth of organs is called organ culture. In this technique the structure and function of the organ is preserved in such a way that it closely resembles the same organ *in vivo*. Animal organs are difficult to study under experimental conditions, because they are not well suited to experimental manipulation and are inaccessible for optical observation. Since preclinical animal studies are limited to investigating human physiology, pathology and therapeutic responses, and traditional human cell cultures which are used to describe species differences are inadequate in their representation of *in vivo* responses, there is immense scope for new approaches. These new approaches include three-dimensional (3D) cultures, organoids, and organs-on-a-chip, which have attempted to better replicate the tissue microenvironment [1]. 2D culture techniques allow researchers to study and manipulate mammalian cells, but 2D cultures do not entirely reflect the 3D structure of cells, as its difficult to achieve or replicate the extracellular matrix within tissues and organs [2, 3]. Thus, previously there has been a huge gap between our understanding of subcellular processes and our knowledge of mammalian biology at the tissue level. The reprogramming/reconstructing of organ structure and function under *ex vivo* conditions can be achieved by 3D, organotypic or organoid cultures. Thus organ culture involves 3D cultures of reconstructed undisaggregated tissue or tissues retaining some or all of the histological features of the tissue *in vivo*. 3D cell culture technology has recently received a lot of attention, as it allows different intensities of cell differentiation and tissue organization, which is not possible in conventional 2D culture systems. 3D cell culture technology is usually carried out using extracellular matrix gels to enhance the expression of differentiated functions and improve tissue organization. This model also faces many shortcomings, such as difficulty in sampling cellular products and probing the effects of physiological diffusion

doi:10.1088/2053-2563/aafac0ch2

gradients [2, 3]. Due to these drawbacks the pharmaceutical industry is still dependent on time-consuming and costly animal studies.

In organ cultures, new growth is in the form of differentiated structures in such a manner that they retain their structural features (e.g. small bronchi in the case of lungs) and retain their physiological features (e.g. hormone-dependent organs remain dependent) in a pattern similar to that in the same organs *in vivo*. In the case of organ culture, outgrowth of isolated cells from the periphery of explants is reduced by controlling the culture conditions. The organ explants should be small since in explants 1 mm or more in diameter there is a chance of central necrosis.

2.1.1 Organ culture techniques

The procedure for *in vitro* organ maintenance was introduced by Fell [4] to investigate bone development. Previously this technique was exploited widely as a morphological technique to study morphological characteristics [5–8]. Currently this method is utilized for many purposes such as for physiological investigations including the influence of vitamins, hormones, carcinogens and embryonic inductors on the physiology of organs. Organ explants may be effectively cultured for up to two weeks under *in vitro* conditions [7, 8]. Unfortunately, a lot of the physical and chemical parameters, e.g. gaseous requirements, pH, osmolality and additional factors, have not been recognized for all organ types. These factors should be well-defined for every specific target organ, keeping in view the age of the donor tissue. Histological studies of explants can be evaluated to determine the architecture of the tissue. This evaluation will offer information about the morphological characteristics of the explant and the cell-to-cell relationships. One of the main merits of organ culture is that it bridges the gap between cell cultures and the *in vivo* conditions, as the structural relationships of the native tissue can be studied.

The organ culture technique was first demonstrated by Loeb (1897). He cultured small pieces of adult rabbit organs (liver, ovary, kidney and thyroid) over a small amount of serum or plasma clots inside test tubes, and noticed that these isolated organs in *ex vivo* conditions retained their normal histological features for three days [9, 10].

In continuation with the previous experiment, in 1936 Parker cultured small tissues derived from an adult organ in a flat bottomed flask containing fluid medium and supplied with 80% oxygen (controlled gas phase). This discovery was followed by several other developments that refined organ culture techniques, as listed in table 2.1.

Cell based model systems have now become a main pillar of drug testing, offering a simple, fast and inexpensive technique to circumvent large-scale and expensive animal testing. As most medical outcomes are based on cellular responses against different factors such as drugs, compounds and external stimuli, cultured cells are considered as an important part of testing treatments. Until now most cell based assays have used traditional 2D monolayer cell culture techniques in which cells are cultured on flat and rigid substrates. Various limitations of this technique have been identified. One of the most important limitations is that under an *in vivo* environment nearly all cells are bordered by other cells and the extracellular matrix in a 3D way, thus 2D cell culture techniques do not offer a sufficiently natural 3D

Table 2.1. Major developments in organ culture.

Development	Researcher	References
Skin slice floating technique (controlled gas phase and shaken flask)	Medawar (1948)	[16]
Embryological watch glass with sealed lid (agar medium)	Wolff and Haffen (1952)	[17]
Watch glass in Petri dish with moist cotton wool (medium in plasma clot)	Fell (1953)	[5]
Watch glass with floating raft of silicon paper	Chen (1954)	[20]
Metallic grid supporting lens paper (controlled gas phase)	Trowell (1954)	[18]
Watch glass with floating raft of silicon paper	Shaffer (1956)	[21]
Watch glass with circular metal grid	Merchant *et al* (1964)	[19]
Organ culture to study human intervertebral disc degeneration	Gantenbein *et al* (2015)	[22]

environment for cells. Therefore, 2D cell culture examinations occasionally offer deceptive and nonproductive information for *in vivo* responses [11, 12]. Currently, the ideal protocol of screening drug molecules is initiated with 2D cell culture assays, which are always followed by animal model tests, before conducting clinical trials. Only around 10% of the molecules successfully pass through clinical development [13–15]. This is due to the information derived from 2D monolayer culture examinations during which the cellular responses against drug(s) may be changed due to the unnatural microenvironment. Thus, due to the lack of an appropriate microenvironment for cell response, many drugs may have failed unecessarily. Recent advancements in animal tissue culture science have allowed the development of new culture techniques, such as 3D cultures, organoids and organs-on-a-chip, which have attempted to replicate efficiently the tissue present in *in vivo*. Table 2.2 encapsulates the advantages of the cell culture techniques being developed.

2.1.2 Plasma clot

Organ culture includes various approaches, of which one of the earliest was the plasma clot method or watch glass technique, introduced by Fell and Robinson in 1929. In the plasma clot method, crude embryo extract is added to a small amount of plasma and allowed to clot. The tissue or organ to be cultured, usually derived from fetal donors, is placed on top of the clot (figure 2.1(A)) and cultured for various time periods to observe temporal changes in morphology and/or cell populations.

The clot consists of a small amount of chick (or other) plasma with chick embryo extract, held in a watch glass. Thus this technique is sometimes referred to as the watch glass technique. This technique is used to culture embryonic organ rudiments [27].

The isolated tissue or organ is cultured for various time periods to study time based changes in morphology and/or cell populations. The watch glass may or may not be closed with a glass lid and sealed with paraffin wax. This has been the classical

Table 2.2. Advantages of human organ culture and emerging cell cultures.

Culture	Advantages
Human organ culture (HOC)	• Offers cell–cell and cell–extracellular matrix interactions while simultaneously preserving biological molecules, in particular proteins, which are essential for growth. • Retains the natural characteristics of the stromal cells or interstitium (the structural part of an organ), native blood capillary networks and parenchymal anatomy (functional part of the organ), allowing healthier responses to external stimuli. • Gene or protein expression, including profiling or imaging, can be assessed on cultures using tissue assays. • Variation between patient-to-patient responses against exogenous factors can be measured.
Organ-on-a-chip	• A microfluidic chamber made up of constantly perfused chambers occupied by living cells, systematically organized to simulate the growth and development of tissues and organs, providing the potential of tissue/organ interconnection. • In contrast to static cultures, smaller media volumes are required. • In contrast to static cultures, the flow of fluid established between channels facilitates extended culture periods and also allows a cell growth environment. • Connecting several organ chips allows scientists to examine organ interactions. It also allows the evaluation of drug efficacy and toxicity predictions. • Assays can be executed in a high throughput and automated mode with a limited number of materials. • The transparent material used for manufacture and the microfluidic nature of the chips, combined with a multi-sensor microsystem, makes imaging simpler.
Organoid culture	• Preserves native tissue characteristics and meticulously replicates 3D structural organization. • Open to high throughput drug screening; 3D organoid models derived from patient cells offer personalized treatment for individuals with different ailments. • A master cell bank can be created by preserving frozen cultures for subsequent use. • Offers the potential to arrest sub-clonal populations under *in vitro* conditions. • Can produce tissue from pluripotent cells. • Can be preserved for more than one year for long-term cultures. • Sequential inspection facilitates the developmental study of tissue.
2D cell culture	• Inexpensive, low cost maintenance, simple to use and easy to manipulate. • Uniformly rich oxygenation and nutrition are provided to all cells, end products are released directly into the medium. • Easier to control, monitor and manipulate the growing cells' micro-environmental parameters.

	• The cultures allow the development of screening platforms for drug discovery, in particular drug testing. • Can easily be monitored and examined using imaging tools
3D cell culture	• Reconstruction of the cell–cell and cell–ECM interface or contact that mimics the *in vivo* environment, encouraging spatial organization and high density culturing. Mimicking the spatial organization of the *in vivo* environment allows variation in responses through manipulating the cell surface receptors involved in interactions with surrounding cells. • Offers a simple approach and is a more efficient technique than 2D culture for studying cellular responses non-invasively and in real time under normal, suitable conditions. • Cells can have natural environment which ultimately increases their life span during prolonged incubation. • Drug testing can be carried out in a more efficient manner, which can offer the probable responses to the drug and any safety concerns for organ-specific cells. • A much better response against external stimuli; tolerates more biotic and abiotic stresses when counteracted by cytotoxic agents.
3D bioprinting	• Offers suitable material with ideal rheological characteristics that can replicate, maintain and improve tissue formation. • Provides extra biocompatibility potential. Also offers additional capacity for targeted cell incorporation at particular sites. • Once ideal conditions are established, offers the potential for high throughput screening.

standard technique for studying morphogenesis in embryonic organ rudiments; it has also been modified to study the action of hormones, vitamins, carcinogens, etc, on adult mammalian tissues.

Plasma clot techniques allow the continued examination of growth without the difficulties of cell traffic (immigration and emigration), which are usually faced *in vivo*. Moreover, it often encourages the maintenance of an organotypic 3D structure under *in vitro* conditions with a complex cellular heterogeneity which is similar to the *in situ* organ.

In detail, the watch glass technique involves placement of a desirable tissue or organ, called the explant, over a clot in a watch glass followed by the transfer of this watch glass into a Petri dish covered with moist filter paper or cotton wool to reduce evaporation of the clot, as illustrated in figure 2.1. After this step the Petri dish is allowed to incubate at 37.5 °C. During this process fresh clots must be provided every 2–3 days for avian tissues and every 3–4 days for mammalian tissues. A modified approach involves the placement of small organ rudiments or pieces on plasma clots (kept on a cover slip) which are then inverted onto the cavity in a cavity microscopic slide (figure 2.1(B)). In this approach the cover slip should be sealed with paraffin wax. Preparation of the clot is done by mixing chicken plasma (three

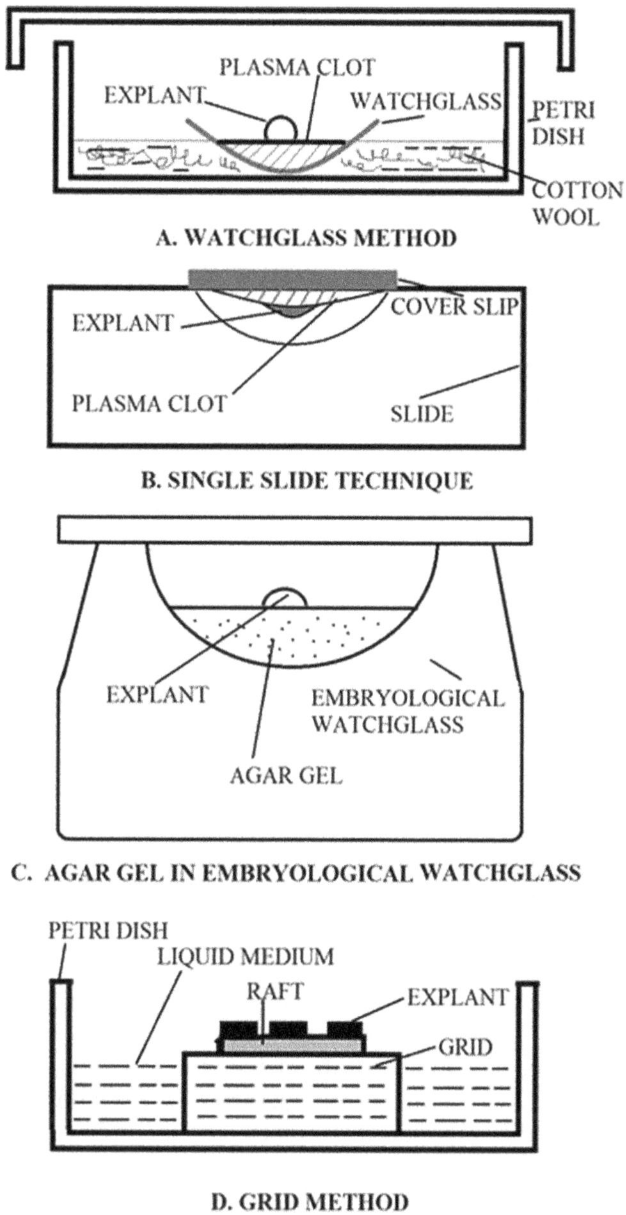

Figure 2.1. Organ culture based techniques. (A) Watch glass method of culture of explants on a plasma clot. (B) Maximow's single slide technique using a cavity slide. (C) Organ culture on agar gel in an embryological watch glass. (D) The grid method contains a metallic wire mesh or a perforated sheet with bent margins; explants are positioned on a raft of lens paper, rayon acetate or millipore filter membrane.

drops) with chick embryo extract (one drop) on the cover slip. In this process the plasma clot can be substituted with fresh clots by simply lifting the cover slip. This method is economical and allows efficient imaging (microscopic observations) during development. Additionally, it is also appropriate for studies such as hair growth, etc. One of the principal drawbacks of the plasma clot approach is that the clot dissolves in the area of the explants so that they become partially or fully immersed in the medium. The life span of a plasma clot culture is somewhat short (less than four weeks) and because of the complexity of the medium biochemical analysis is not possible.

2.1.3 Raft methods

Raft methods ultilize lens paper or rayon acetate (water resistant), which is used for cleaning optical lenses (figure 2.1(D)). This paper/fabric is called a raft because it can float on the fluid medium. Chen (1954) demonstrated the utilization of lens paper in the form of a raft to position cultures so that they float over the fluid medium. He transferred four to five explants of an isolated tissue onto a raft of lens paper (25 × 25 mm) floated on serum in a watch glass. Shaffer (1956) replaced the lens paper with fabric, rayon acetate. Rafts of rayon acetate are prepared to float on the serum by treating them with silicone, which will also enhance the floatability of the lens paper [23]. The idea of treatment with silicone was demonstrated by Richter (1958), who suggested that any sort of lens paper will float if it is first treated with silicone. Rayon acetate rafts have the advantage over lens paper of being acetone soluble, thus they can easily be solubilized in acetone for histological assessments. On an individual raft, four or more cultures are generally maintained. The raft modification permits access to any changes in the medium and also averts the dipping of cultures into liquefied plasma. In 1990, Barabonov modified the Chen raft technique by suggesting the use of rafts of hydrophobic polymers (polyethylene) [23]. Rafts were prepared as disks with a hole in the middle. The prepared rafts were reusable, with good buoyancy, and remained on the surface of the culture medium for more than one week. Similar to the plasma clot procedure, the raft system permits the explant to be kept at the gas–medium interface in comparatively well-controlled *in vitro* conditions. Moreover, raft techniques provide better control of soluble factors through the use of defined media and growth factors, if necessary [24, 25]. Some limitations of the raft method are that no more than four small cultures can be placed on the raft, otherwise it may sink and becomes increasingly difficult to handle. Also, when the medium has to be changed the raft must be lifted off, which always presents difficulty in terms of handling [24, 25]. Chen observed that, to avoid the paper sinking, the top surface should not be allowed to become wet. Additionally, Shafer also noticed that protein spilling from the medium over the silicone–rayon paper can restrict its floating, which can ultimately affect the whole experiment [24, 25]. These shortcomings can be overcome by monitoring the wetness, so that it can be controlled in a proper way, however it is somewhat impractical to control wetness, thus the grid method was introduced which provides complete control over the wetness and is much easier to handle.

2.1.4 Grid method

This method was introduced by Trowell in 1954 [18], in which instead of utilizing paper a metallic grid is utilized to support the cultures over the liquid medium (figure 2.1(D)).

The grid is square with dimensions of 25 mm × 25 mm. The grid stands in a shallow dish and sufficient medium is added to wet the top of the surface. These grids were previously fabricated with tantalum wire gauze (often used in surgery) but due to the rigidity of this metal, now expanded metals, which should be nontoxic and non-corrosive, are used instead of wire gauze. In this method the tissues or cultures are usually placed directly over the grid, whereas softer tissues glands or skin are initially placed on rafts, which are then kept on the grids. The grids themselves are in a sealed culture chamber (which consists of a Perspex Trowell Type I, with a small gas volume and gas remaining up to 24 h, appropriate for short-term investigations, or a Trowell Type II chamber, made of aluminum, with a large gas volume and gas remaining up to 3 days, for long experiments) filled with fluid medium up to the grid, and the chamber is supplied with a controlled gas supply of O_2 and CO_2 to meet the high O_2 requirement of adult mammalian organs [24]. In 1959, Frank demonstrated the utilization of a Petri dish to hold the culture (including cultures which do not have high oxygen demands), which can be utilized to study the growth and differentiation of adult embryonic tissues. Later, in 1961, Prop demonstrated the Trowell Type II chamber in the form of a screw capped jar [24].

2.1.5 Agar gel

This procedure was discovered by Spratt in 1947 [26]. The agar gel method is now extensively used, mainly for the culture of embryonic organs (figure 2.1(C)). This procedure is similar to the watch glass technique and it is successfully used for developmental and morphogenetic studies.

In his earlier experiments, Spratt used a nutritive medium containing salt, serum, chick embryo extract, white egg yolk or albumin alone mixed with 1% solution of agar, which is used as the substrate. Later, in 1948, he introduced a synthetic medium which contains a mixture of certain amino acids, vitamins and sugar, mixed with a 1% solution of agar. This procedure prevents the sinking of tissues into the medium and allows the use of a defined medium. Usually, isolated tissues must be subcultured within an interval of 5–7 days on fresh agar gels. The agar gels are usually kept in embryological watch glasses and sealed with paraffin wax. The cultured tissues can be studied by means of a stereoscopic microscope. This approach can be used to investigate many growth phases of normal organs as well as of tumors.

2.1.6 Cyclic exposure to the medium and gas phase

Using this procedure, various human adult tissue culture (esophagus, mammary epithelium, uterine endocervix, etc) can be maintained for longer periods, i.e. up to 4–5 months (figure 2.1(D)). In this procedure isolated tissues are occasionally exposed to the fluid medium and the gas phase [27]. The isolated tissues number varies from 2–18, based on the organ cultured. These isolated tissues are attached to the bottom of a plastic culture dish and then concealed with fluid medium. Then the dishes are placed in a chamber supplied with a suitable gas mixture in a controlled manner [27]. The chamber is mounted on a continuously agitated rocker platform and is agitated at desirable speed to ensure the cyclic exposure of the isolated tissue to the medium and gas phases.

2.1.6.1 Advantages

Organ cultures have the following advantages:

- Cultures are exposed to an environment which is more or less comparable to *in vivo* conditions in both structure and function. This creates more a suitable environment than cell cultures for physiological investigations.
- The development of fetal organs *in vitro* is comparable to that *in vivo*. Hormone-dependent organs remain so, while endocrine organs secrete specific hormones.
- Organ cultures offer information about developmental patterns, differentiation and growth.
- It has been recently discovered that organ cultures can replace entire animals in animal based experimentation, because results obtained from organ cultures can be easily understood.

2.1.6.2 Limitations

Organ cultures suffer from the following limitations:

- The similarity between the results obtained from organ cultures and those obtained from *in vivo* environments can be low. For instance, while studying drug action, it has been observed that drugs are metabolized *in vivo* but not under *in vitro* conditions.
- Organ cultures can only be preserved for short periods of time (a few months), however, in some cases it is desirable to investigate the effects of certain factors for longer. In such cases cultured organs should be transplanted into suitable host animals, e.g. nude mice.
- They are not suitable for biochemical or molecular investigations.
- For each and every experimental examination, fresh organ cultures have to be developed.
- They can be studied using a limited number of studies, such as histological, immune-cytochemical and auto-radiographic studies *in situ*.

Thus, the results obtained from organ cultures generally offer information on *in vivo* events; this frequently decreases significantly the total number of experiments required to examine a particular problem. Organ culture allows us to preserve the original structural–functional relationship between different cell species and also allow us to examine the long-term effects of external stimuli.

2.1.6.3 Applications

Organ cultures have been widely used for the following:

- *Growth and development related studies.* Investigation of the patterns of growth, differentiation and development of organ rudiments, i.e. fetal organs, and the effects of different elements, e.g. hormones, vitamins, etc, on these parameters.

- *Drug action.* The action of drugs (e.g. cytotoxic drugs) on the organs can be investigated under *in vitro* conditions. This investigation of drugs can be considered as a guide for the events in whole animals.
- *Tissue engineering.* Organ culture offers a foundation to produce tissues for implantation in patients, e.g. human skin can be successfully produced *in vitro*. Ultimately, it is helpful in the reconstitution of organs *in vitro,* which can be further utilized as grafts or transplants and models for studies on drug delivery and action. Moreover, such reconstitution allows the development of skin, cartilage, liver and pancreas, which can be used in organ transplantation for patients with serious health conditions.
- *3D environment.* 3D matrices are now extensively utilized in tissue engineering. They are available in different forms, such as inorganic (calcium phosphate) and organic (collagen gel, cellulose sponge (either alone or coated with collagen) and Gelfoam). This type of environment allows the growth of anchorage-dependent cells. These matrices encourage cell attachment, proliferation and differentiation. Moreover, they act as scaffolds and help cells organize themselves in a 3D structure. Thus they offer a platform in the form of a gel matrix which allows cell growth either at the solid–gel interface or within the gel itself. Cells cultured over this matrix can be used for several therapeutic purposes. Thus, these matrices are fabricated with such materials that they can easily be degraded under *in vivo* conditions, which can ultimately result in the development of an endogenous matrix [28–31]. Plasma clots or fibrin clots are considered as the first form of 3D matrix, and are still used for primary cultures and, in the form of plasma clots or purified fibrinogen mixed with thrombin, for numerous applications. For example, an autologous plasma clot matrix has recently been reported, which possesses a microporous 3D structure with outstanding biocompatibility and also enables the diffusion of nutritional and regulatory factors [28–31]. The additional benefit of an autologous plasma clot is that it does not require allogeneic or xenogeneic factors during clot preparation, in contrast to commercially available fibrin sealant systems [28–31].

2.2 Histotypic culture

The development of cell lines with high cell density in a 3D matrix is called a histotypic culture. This approach of culturing cells offers the unique advantage that dispersed monolayer cultures can be utilized to regenerate tissue-like structures. The most common techniques in histotypic culture are the gel, sponge and hollow fiber techniques. A histotypic culture is a 3D culture of one cell type, whereas organotypic suggests the interaction of two or more cell types from a complex tissue or organ. In the histotypic technique once the cells are attached to capillary fibers they grow, which will result in the increase of cell density, forming tissue-like structures. Several researchers have reported that this type of structure is comparable to the tissue present in *in vivo* behavior. This culture technique is considered as the perfect model for the commercial production of several biologically significant compounds.

2.2.1 Gel and sponge technique

To replicate the *in vivo* environment for cells, gels and sponges are considered the best approaches. They are purified extracellular molecules and biopolymers. Gel biopolymers are poured by the user or purchased precast in the culture flask or assay plate, and the most common are gelatin, collagen and laminin. To mimic cellular microenvironments, sponges such as lyophilized AlgiMatrix™ gels with large pores can be utilized. Different substrates have emerged at the commercial level, such as Matrigel™ (BD Biosciences), Extracel™ (Glycosan Biosciences) and AlgiMatrix™ (Invitrogen) (table 2.3).

Leighton initially introduced this technique by demonstrating that the normal and cancerous cells can penetrate cellulose sponges which offer a matrix for the development of primitive cells [32, 33]. Later, Gelfoam, which is made up of gelatin sponge, has been utilized in place of cellulose [34]. All these models require histological investigations and are inadequate in their dimensions. Utilization of 3D sponges and gels has increased tremendously with advancements in tissue engineering. This procedure can be utilized to develop mammillary epithelium, and tubular and glandular structures. Collagen gel (native collagen) and Matrigel (a commercialized product made up of laminin, collagen, fibronectin and proteoglycans, with or without growth factors) can be utilized in this technique.

2.2.2 Hollow fiber technique

To achieve a high density of cells and to retain their structural integrity and functional properties, many designs have been created to mimic the *in vivo* environment, including plastic capillary fibers, called hollow fibers (figure 2.2). Cells are cultured over the fibers and nutrients are supplied via capillary spaces. Since medium supply and exchange are not adequate at high cell densities, in this approach a perfusion chamber with a bed of plastic capillaries can be utilized to encourage nutrient and gas exchange in a more efficient way [35]. This design is suitable for studying the synthesis and release of certain biochemicals, such as cytokine and monoclonal antibodies.

2.2.3 Spheroids

A spheroid is a group of cells, usually formed by the reassociation of dissociated cultured cells. The basic principle involved here is the reunion or sorting of mainly heterotypic (or sometimes homotypic) cells to form tissue-like structure. Spheroids can be broadly divided into two categories:

- Spheroids derived from reaggregation of monolayer cells of cancer cell lines are homotypic. In this case similar types of cells form the spheroid.
- In contrast, when mixed types of cells from the cell lines of different lineages reaggregate to form spheroids they are heterotypic.

Table 2.3. Types of substrates commercialized to replicate the *in vivo* environment.

Type of matrix	Source	Composition	Function	Favorable for
Matrigel™	Derived from Engelbreth–Holm–Swarm tumor grown in mice	Composed of IV collagen, laminin and heparin sulfate.	Acts as a basement membrane.	Epithelial cell culture
Extracel™	Hyaluronan based gel which creates a compressible hydrogel similar to the structure of a joint	Made by combining hyaluronan, gelatin and the crosslinker polyethylene glycol diacrylate. This defined composition can be modified by the addition of other extracellular components, such as laminin, and the gel stiffness adjusted by the fixation crosslinking procedure.	Offers some extracellular matrix components for cell attachment.	3D cell culture
AlgiMatrix™	Animal-free product derived from brown seaweed that gels in the presence of divalent cations to form a negatively charged hydrogel	A ready-to-use sponge made from lyophilized alginate gel.	Allows cells to invade the pores and secrete endogenous ECM components that support *in vivo*-like morphologies, structures and behaviors.	Primary and stem cell spheroid culture

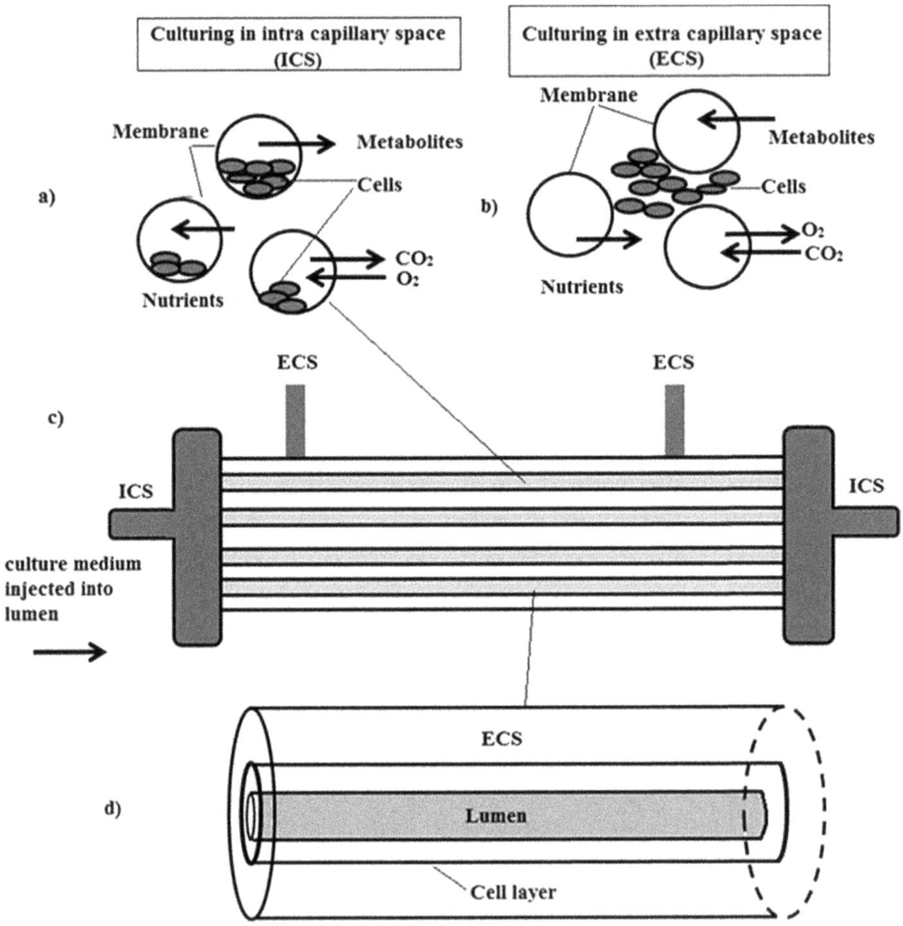

Figure 2.2. Hollow fiber technique: (a) culturing in intracapillary space; (b) culturing in extra capillary space; (c) culturing with hollow fiber bioreactors; and (d) construction of hollow fiber reactors.

Heterotypic spheroids can be derived from freshly trypsisnized cancer tissue which is related to vascular micro-metastasis. Spheroids are broadly experimented on to investigate the following:

- Penetration of drugs.
- Resistance exhibited by a particular tumor against cytotoxic drugs.
- The response of a cancer cell to radiation.
- Susceptibility to drugs.
- To investigate the effect of monoclonal antibodies tagged with toxins, prodrugs and radionuclides in targeted therapy.

Moreover, sometimes cancer and normal cells independently form spheroids in the same culture, which are called mixed spheroids, and can be used to investigate the invasive properties of cancer cells. Heterotypic spheroids can be considered as *in vitro* models for interpreting cellular interactions, in particular in inflammatory diseases, diabetes and complex diseases.

Spheroids suffer from major limitations related to diffusion and exchange of nutrients and gases. Once dispersed cells are grown in a gyratory shaker, they may reunite to form clusters. This reassociation of disaggregated cells, especially in the case of embryonic tissues, is achieved in a unique way [36]. Cells in these different aggregates are capable of arranging themselves into clumps or aggregates and form tissue-like structures. These are considered as systems for chemotherapy *in vitro* [37] and for the characterization of malignant invasion [38]. Like organ culture, the development of spheroids is inadequately nourished by diffusion, and over a period of time a stable state may be reached in which cell multiplication is stabilized by central necrosis.

2.2.3.1 Multicellular tumor spheroids

Multicellular tumor spheroids offer an *in vitro* proliferating system for the investigation of cancerous cells. The 3D arrangement of multicellular tumor spheroids enables studies related to drug testing, in particular drug penetration. The main benefit of this model is that it offers a well-defined planar or spherical arrangement of cells which is comparable to the structure and function of the tumor *in vivo*. However, beyond a critical size most of the multicellular tumor spheroids develop necrosis at their center, surrounded by viable cells (figure 2.3). Transfectant mosaic spheroids can also be cultured from cells that have been transfected with different genes. In this process cells are initially cultured as a monolayer and later they are transfected with the transgene and then considered for selection for transgene-expressing cells. The addition of transfected and nontransfected monolayer cells in any desired amount can result in the formation of mosaic spheroids [39, 40].

2.2.4 Rotating chambers

2.2.4.1 Miniperm bioreactor

In this process continuous agitation and aeration can be achieve by rolling the culture vessel, either in a conventional roller bottle or in two-compartment chambers

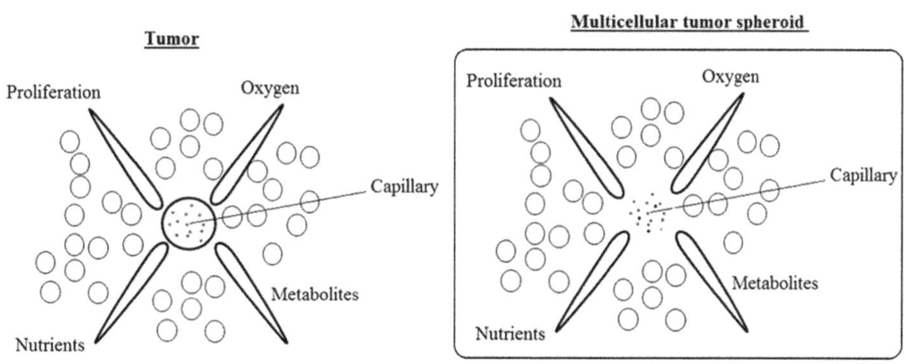

Figure 2.3. Tumor in comparison to a multicellular tumor spheroid.

(a cell chamber and medium chamber). To achieve a high cell concentration, the cell suspension is allocated to one small compartment, as this can avoid the dilution of cells by the bulk of the medium. In the cell compartment, a product such as an antibody will accumulate, whereas nutrients and other waste products move from the semipermeable membrane to the medium compartment (figure 2.4). Figure 2.4 shows a unique design of a miniperm bioreactor, which contains two separate compartments. The smaller compartment contains the cells/product and the larger compartment contains the medium. These two compartments are separated by a semipermeable membrane which encourages the proper mixing and separation of products from the medium. The design of this chamber and the slow rotation allows aggregate formation, which may ultimately enhance product formation.

2.2.4.2 Rotatory cell culture system (RCCS)

In the late 1980s, to attain unobstructed 3D growth in suspension on Earth, NASA developed a rotating wall vessel in which cells, growing in suspension, attained simulated zero gravity (as occurs in microgravity) with a gently rotating chamber changing the sedimentation vector uninterruptedly [41–44] (figure 2.5). During this process cells in the stationary phase are exposed to zero shear force. These cells further form 3D aggregates called spheroid-like structures. Gas can be supplied from a central silicone membrane to the cell-containing cylinder. When revolution halts aggregates form and sediment settles. During this time the medium can be replaced. This design was created to protect cell culture experiments from high forces. The architecture of this design encourages tissue development and thus can be used in tissue engineering laboratories on Earth. This bioreactor consists of a cylindrical vessel capable of rotating at such a constant speed that equal forces are achieved between the downward gravitational force and the upward hydrodynamic drag force acting upon each tissue fragment [45].

2.2.4.3 Cell based immobilization

The approach of cell immobilization is considered as a vital method in current biotechnology [46–49] as it not only offers enhanced fermentation productivity, but

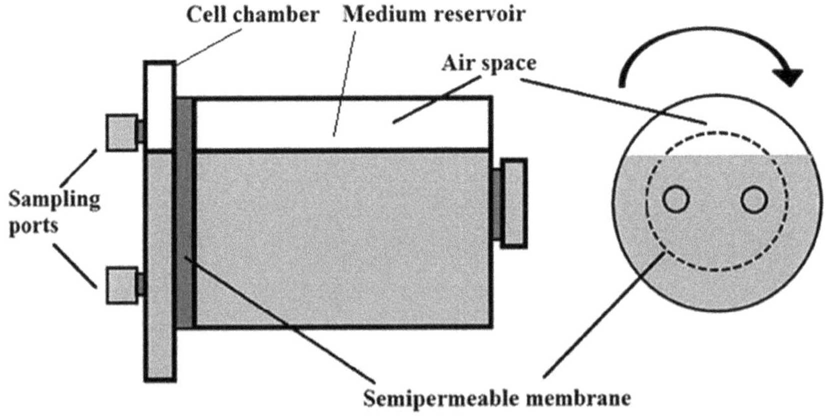

Figure 2.4. Rotating chamber system.

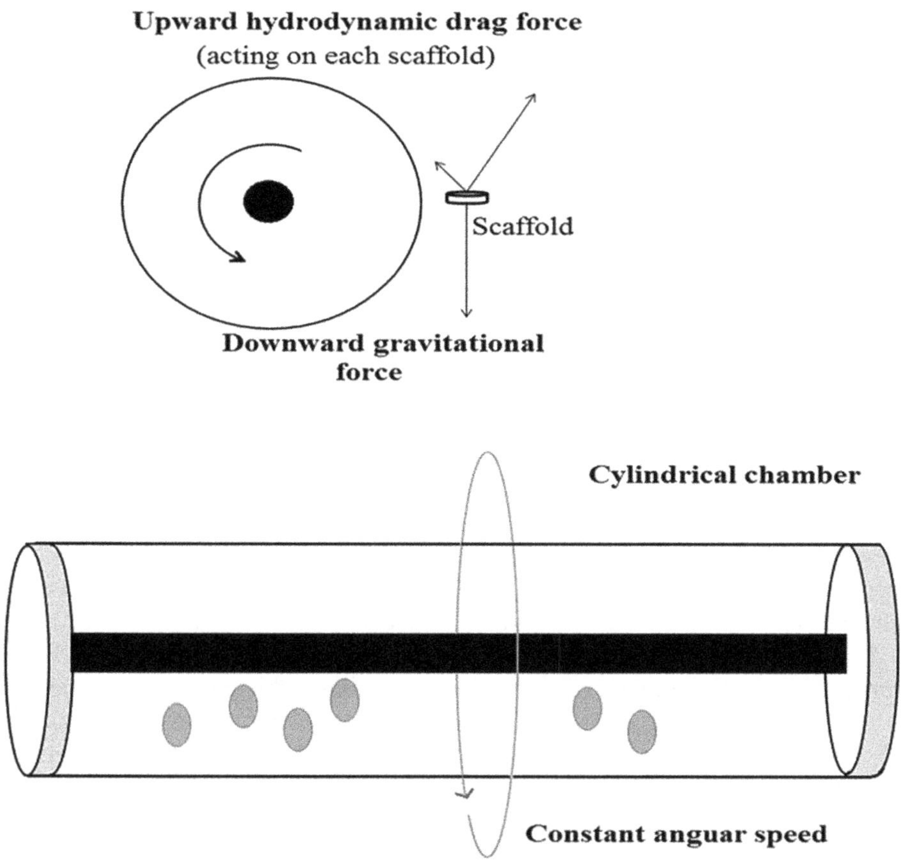

Figure 2.5. Rotating wall chamber developed by NASA.

also the feasibility of continuous processing, high cell stability and low cost of recovery [50–53]. Encapsulation of living cells within alginate beads has been widely used in experimental research for product formation. Recently, immobilization of cells within spheres of Ca^{2+} alginate has extensively been utilized for immobilizing living cells [54]. For example, monoclonal antibody production from hybridoma cells, hormone-producing cells used in animal models for the treatment of diabetes mellitus [55] and chondrocyte differentiation [56]. In one study yeast cells were immobilized by initially coating them with polyelectrolytes (i.e. microcontact printing of polyelectrolytes) and then immobilizing them over the patterned surfaces. By using this approach the coated cells are shielded and their cell–substrate interactions are modified [57].

Recent developments have explored an array of support materials for cell immobilization, such as inorganic materials [58], organic supports [59] and natural supports [60]. In contrast to other substrates, inorganic mesoporous material is useful as it offers a large specific surface area, a narrow pore distribution, and good resistance against biodegradation [61]. Additionally, its low toxicity, low cost and excellent biocompatibility have allowed scientists to utilize it for cell immobilization

[62, 63]. However, inorganic mesoporous material has rarely been used as a support material to immobilize *living* cells due to its requirements for active conditions such as extreme pH and high temperature and pressure throughout the immobilization process [64, 65]. Thus, utilization of inorganic mesoporous materials to immobilized living cells under mild conditions is a challenge.

Thus novel approaches are required for cell immobilization. Recently, a newly architecture model called 'fish-in-net' has been used for the encapsulation of enzymes in ordered mesoporous silica under mild conditions with the main focus being to prove the potential of this strategy in the immobilization of living cells [66].

2.2.4.4 Filter well inserts

Because of their better cellular interaction, stratification (formation or deposition of layers) and polarization in culture systems, filter well inserts (which hold microporous filter membranes) have become popular for organ cultures and have been used in various forms. They also offer a substrate for cell attachment over the basement membrane, better reaggreagation of cells to form tissues, and enhanced cell density, access to the medium and gas exchange. Simultaneously, these inserts support the secretion of various molecules from the basal and apical surfaces (figure 2.6). Filter well inserts are considered as one of the first technologies that initiated the approach of 3D-like exposure of cells to a substrate by permitting all sides of the membrane to interact with the environment. This *in vivo* mimicking architecture allows the study of both surfaces of a cell monolayer, such as epithelial cell line migration, development and tissue modeling, as filter cultures allow the generation of stratified epidermis. In addition, this 3D structure allows directional polarized metabolic processes similar to those that occur *in vivo*. Polarization and functional reliability can be established, for example in thyroid, intestinal and kidney epithelium. Filter well inserts are available in different designs, sizes, coatings and pore sizes. Millicell (by Millipore) and Transwell (by Corning) are considered as the two major commercially produced filter well inserts (figure 2.6). The first is made up of four membranes (a biopore membrane, MF-Millipore™ membrane, Isopore™ membrane and polyethylene terephthalate), whereas the latter is made up of Transwell-COL inserts made of collagen-treated PTFE with a trademarked collagen coating to increase cell adhesion.

Filters coated with collagen, fibronectin, metrigel and laminin are available. These filter inserts are available in different sizes. Currently, filter wells are available in the different forms shown in figure 2.6. One of the major advantages of filter well inserts is that they allow the culture of cells at very high, tissue-like densities, with ready access to medium and gas exchange, but in a multireplicate form. They also allow recombination of multiple different cell types with or without a matrix. Filter well inserts are now available from several suppliers in a variety of translucent or transparent materials, including polycarbonate, PTFE and polyethylene teraphthalate, and ranging in size from 6.5 mm to 9 cm, suitable for 6, 12 and 24 well plates, or larger dishes.

Recently, a 3D co-culture system with filter well inserts was developed with both adipose-derived stem cells and human dermal fibroblasts to study the influence of dermal fibroblasts on the expression of keratinocyte markers in adipose-derived stem

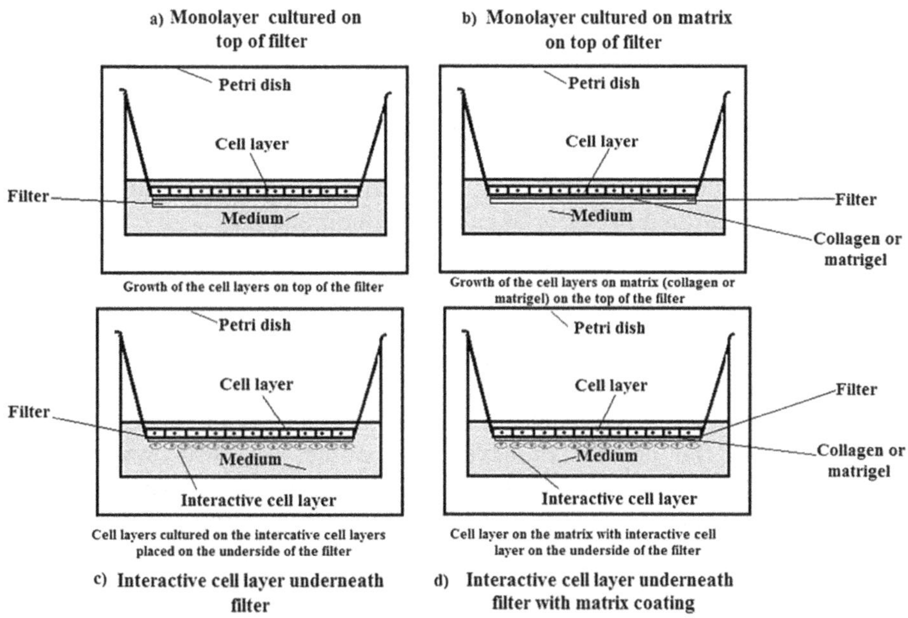

Figure 2.6. Illustration of organ culture in filter well inserts.

cells, in particular during the optimization of the *in vitro* system at the air–liquid interface [67]. However, this process was not suitable for use in clinical applications. Filter well inserts have been successfully used to develop functionally integrated thyroid epithelium [68], stratified epidermis [69–78], intestinal epithelium [69–78], renal epithelium and granulocytes or malignant cells [69–78]. In addition, this system can also be used to study cell invasion [69–78]. Others have used them to study invasion by granulocytes or malignant cells.

2.2.4.5 Cultures of neuronal aggregates

Aggregating cultures of fetal brain cells have been extensively used to study neural cell differentiation [78–83]. The aggregating cells follow the same developmental sequence as observed *in vivo*, leading to an organoid structure consisting of mature neurons, astrocytes and oligodendrocytes. A prominent neuropil is also formed. In tumor biology, the aggregates can be used to study brain tumor cell invasion *in vitro* [83].

2.3 Artificial skin

Communication between cells, and between a cell and its surroundings and extracellular matrix, is required for the control of cell behavior, which will ultimately control the growth and development of the cell at particular times [84–89]. Thus cells cultured in 2D monolayers cannot capture the significant complexity of *in vivo* conditions [90]. For example, cells experience the loss of numerous important signals, tissue phonotypical characteristics and key regulators when they are allowed to develop on 2D substrates such as culture plates. When cell growth takes place in a 3D environment it allows the expression of different cell surface receptors, proliferative capability, extracellular matrix synthesis, variation in cell density and metabolic functions [91]. The 2D method for culturing skin cells not only fails to develop the complex and active environments of *in vivo* tissues, but can also produce incorrect results by compelling cells to adjust to an artificial, flat and rigid surface. There are a number of reports that show the variation in the phenotypical characteristics, cellular signaling, migration and response to different drugs when similar cells are cultured in 2D or 3D culture conditions [92].

Skin is considered as the largest organ of the human body. Anatomically it is arranged into an elaborate layered structure containing primarily the outermost epidermis and the basic dermis. Due the developments in animal tissue culture science, it is now possible to produce skin *in vitro*. Previously, skin developed *in vitro* has only been the epidermis portion of the skin, which can be utilized to replace skin in a burnt area, which can ultimately result in the regeneration of the dermis. New developments in the procedures have allowed the reconstruction of virtually complete skin (both epidermis and dermis). This procedure produces what is called living skin equivalent, and utilizes a collagen matrix as a support for growth of the tissue. Living skin equivalent is a 3D organotypic model, and has been extensively utilized to study numerous features of cutaneous biology. Skin tissue can be derived from the patient, from a foreskin (the loose skin at the tip of the penis) or from newborn babies. However, there are comparatively few investigations evaluating how authentically the skin equivalent reproduces normal skin biology. Skin equivalent is made by culturing human epidermal keratinocytes over a hydrated collagen lattice populated with human dermal fibroblasts, which is then raised to the air–liquid interface where keratinocyte development results in the formation of tissue [93, 94]. This presents numerous common morphological characteristics of normal skin. A comparison between normal and artificially grown skin is shown in figure 2.7.

It is well known that the skin cells of a newly born baby develop more strongly than mature or adult skin, and it has been reported that the utilization of a synthetic polymer such as PGA permits newborn skin to develop without scars. Newborn-baby-derived artificial skin explants can be utilized to seal and cover a wound until the patient's skin is cultured and artificial skin is available for grafting. The *in vitro* development of artificial skin begins with the preparation of the dermal equivalent, consisting of type I collagen and fibroblasts. Once they are polymerized, the plating with keratinocytes and melanocytes is carried out. After 24 h plating and once the collagen gel has contracted, the whole arrangement is passed from a steel grid to finally form the air–liquid interface. Formation of artificial skin takes place on the interface after 2 weeks, as shown in figure 2.8. Another approach is the culturing of keratinocytes over a feeder layer consisting of irradiated 3T3 fibroblast cell line. The epidermis is made up of keratinocytes, which produce the dead cells called corneocytes, which will ultimately form the outermost cornified layers of skin. Skin explant treatment with trypsin is required to disaggregate keratinocytes. Later, these dissociated cells are cultured in culture vessels the extremity of which is concealed by a feeder layer comprised of irradiated 3T3 fibroblast cell line. These fibroblast cells allow the proliferation of keratinocytes. During this event

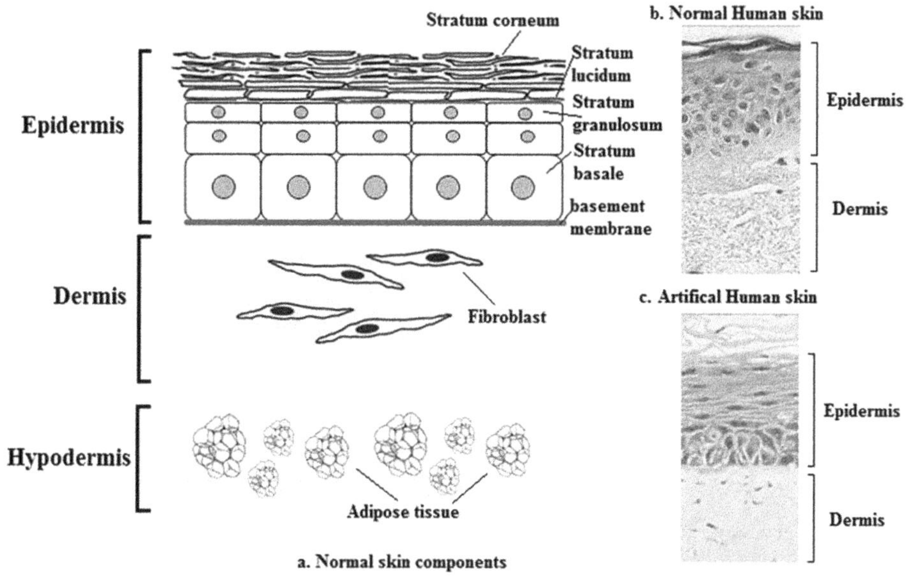

Figure 2.7. (a) Normal anatomical features of skin. (b) Normal human skin (optical microscopical features). (c) Artificial human skin (optical microscopical features).

keratinocytes develop into colonies and are separated into single cells and cultured in a similar way. This entire procedure is repeated again until a confluent stage of pure epithelium is achieved. This can be achieved by using suitable matrices that support the 3D growth of cells. The graft is obtained by detachment of the tissue-sheet along with the matrix from the culture vessels, which is then cleaned and ultimately used for grafting. The tissue used for the culture must be obtained from the patient to avoid any sort of rejection. As per reports, using this technique within 3–4 weeks an explant of size 3 cm^2 can develop into 1.7 m^2 of artificial skin. This represents a 5000-fold increase in tissue size which is significant. After five years, regeneration takes place and grafted skin develops all the vital components of skin. Artificial skin grafts have been utilized to effectively regenerate numerous types of skin scars and chronic skin ulcers. Based on the review by Halim *et al* (2010), skin substitutes (biological and synthetic) play a significant role in the management of different skin injuries, however, due to our lack of understanding at the molecular level, at present there is no perfect substitute available on the market. Synthetic skin substitutes are those substitutes which can be synthesized as required and can be modified for specific purposes, whereas biological skin substitutes have a more intact extracellular matrix structure [95]. Each class has its own advantages and disadvantages. Biological skin substitutes encourage building of a more natural new dermis and basement membrane and allow excellent re-epithelialisation, while synthetic skin substitutes provide the benefits of increased control over support composition [95].

Due to the unavailability of artificial skin that entirely imitates normal uninjured skin, natural biopolymers, e.g. collagen and fibronectin, are being considered as possible sources of biomaterial to which cells can attach [96]. Engineered skin

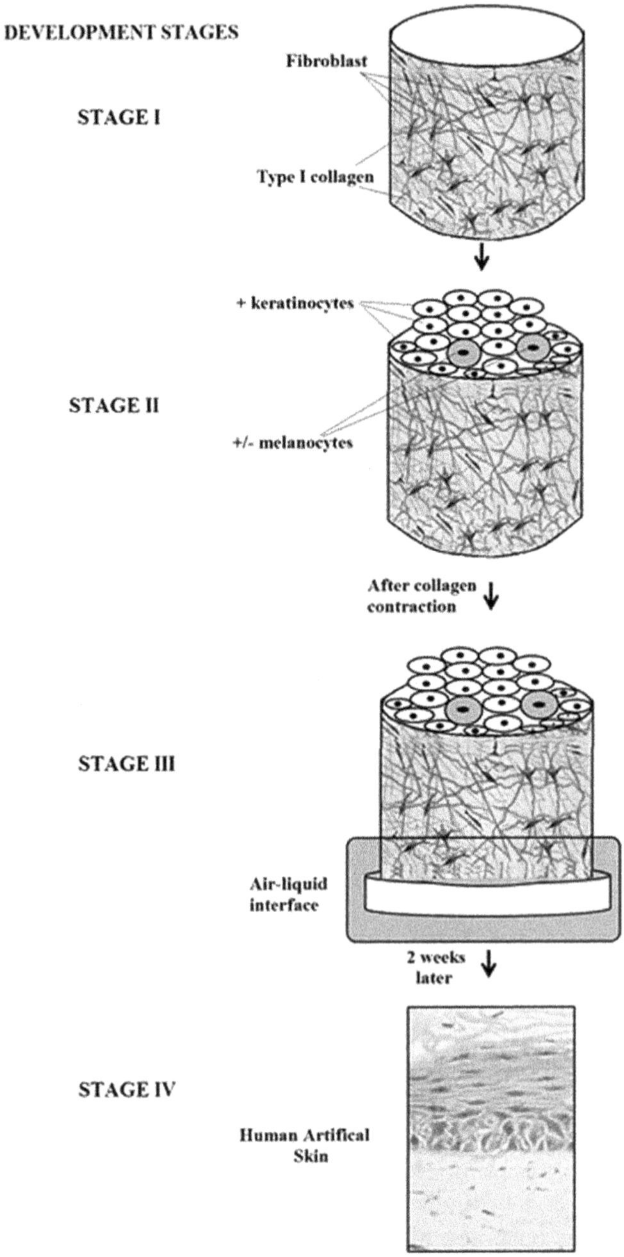

Figure 2.8. Development of artificial skin.

substitutes are developed from acellular materials or can be synthesized from autologous, allograft, xenogenic or synthetic sources [97].

The earlier degradable polymers employed in tissue engineering have considerable disadvantages in terms of mechanical and degradation properties, which resulted in the development of synthetic degradable gels. These gels are mainly utilized to deliver cells and/or molecules *in situ*. This technology is called smart matrix technology. The physiological repair of tissues or organs is accompanied by fibrotic reactions which can lead to the production of a scar, however, there are several organisms in nature with the capability to regenerate without scarring, e.g embryonic or fetal skin and the ear of the MRL/MpJ mouse [97]. Studies of such models to clarify our understanding of regenerative processes have proved that the cascade of inflammatory responses is different in such a way that the scope for fibrosis and scarring is much reduced. One of the major obstacles in investigating the mammalian regenerative process is to explore the number of factors and cytokines expressed during regeneration and incorporate them to create a smart matrix for use in a skin equivalent. The challenge is to identify the factors that influence regeneration, and incorporate these factors to make a smart matrix for use in a skin equivalent, e.g. cytokines expressed during regeneration [97]. Current advancements in the techniques available in genetic engineering allow the use of DNA microarrays and proteomics to aid in the identification of such molecules. Additionally, technologies such as non-viral gene delivery and stem cell technologies can also contribute to skin replacements [97].

The challenges of the easy rupture or tear of artificial skin made of collagen gel during suturing and grafting was overcome by fabricating artificial skin comprised of a stratified layer of keratinocytes and a dermal matrix with a type I collagen containing fibroblasts, i.e. a histocompatible collagen mesh was prepared and attached to the bottom of the gel [98].

Additional challenges such as immunological rejection of skin substitutes are very common. Most skin substitutes are still vulnerable to immunological rejection, and currently no skin substitute is available that has overcome this phenomenon [99].

Recent research has explored an *in vitro* co-culture system which contains mammalian cells and marine green algae, *Chlorococcum littorale,* to make a 3D tissue. In this study muscle cells derived from a mouse and cardiac cells from a rat were studied. It was observed that cardiac cells obtained from the rat consumed oxygen actively. Moreover, the introduction of this co-culture system changed the anaerobic respiration in cardiac respiration to aerobic respiration [101–122]. This co-culture system significantly decreased the consumption rate of glucose, the production rate of lactate and the total concentration of ammonia. Cells cultured without algal components were also investigated and it was observed that most of the cells suffered from delamination and other cellular injuries due to the release of creatine kinase. In contrast, algae co-cultured cells maintain a good histological state. This recent finding by Haraguchi *et al* opened the way for other algae to be evaluated for their potential co-culture with different animal cells [100]. There are several algae available in the vast diversity of species which contain vital amounts of secondary metabolites which can be utilized to develop suitable co-culture candidates. One of the algae is *Porphyra vietnamnesis*, which contains various secondary metabolites and possibly can be utilized in co-culture systems [101–122]. This co-culture system can be utilized to improve the culture conditions to further encourage 'a symbiotic recycling system' composed of mammalian cells and algae, as suggested by Haraguchi *et al* [100].

References

[1] Al-Lamki R S, Bradley J R and Pober J S 2017 Human organ culture: updating the approach to bridge the gap from *in vitro* to *in vivo* in inflammation, cancer, and stem cell biology *Front. Med.* **4** 148

[2] Huh D, Hamilton G A and Ingber D E 2011 From three-dimensional cell culture to organs-on-chips *Trends Cell Biol.* **21** 745–54

[3] Shamir E R and Ewald A J 2014 Three-dimensional organotypic culture: experimental models of mammalian biology and disease *Nat. Rev. Mol. Cell Biol.* **15** 647–64

[4] Fell H B and Robison R 1929 The growth, development and phosphatase activity of embryonic avian femora and limb-buds cultivated *in vitro Biochem. J.* **23** 767–84

[5] Fell H B 1953 Recent advances in organ culture *Sci. Prog.* **162** 212–31

[6] Kahn R H 1958 Organ culture in experimental biology *Med. Bull. Ann. Arbor.* **24** 242–52

[7] Dawe C J 1963 *Symposium on Organ CultureNational Cancer Institute Monographs* vol 11 (Washington, DC: US Department of Health, Education, and Welfare)

[8] Thomas J A 1970 *Organ Culture* (New York: Academic)

[9] Loeb L 1897 *Über die Entstehung von Bindegewebe, Leueocyten und roten Blutkörperchenaus Epithel und übereinemetode, isolierte Gewebsteilezuzüchten* (Chicago, IL: Stern)

[10] Loeb L and Fleisher M S 1919 The growth of tissues in the test tube under experimentally varied conditions with special reference to mitotic cell proliferation *J. Med. Res.* **40** 509–50

[11] Birgersdotter A, Sandberg R and Ernberg I 2005 Gene expression perturbation *in vitro*—a growing case for three-dimensional (3D) culture systems *Semin. Cancer Biol.* **15** 405–12

[12] Weaver V M *et al* 1997 Reversion of the malignant phenotype of human breast cells in three-dimensional culture and *in vivo* by integrin blocking antibodies *J. Cell Biol.* **137** 231–45

[13] Bhadriraju K and Chen C S 2002 Engineering cellular microenvironments to improve cell-based drug testing *Drug Discov. Today* **7** 612–20

[14] DiMasi J A and Grabowski H G 2007 Economics of new oncology drug development *J. Clin. Oncol.* **25** 209–16

[15] Breslin S and O'Driscoll L 2013 Three-dimensional cell culture: the missing link in drug discovery *Drug Discov. Today* **18** 240–9

[16] Medawar P B 1948 Tests by tissue culture methods on the nature of immunity to transplanted skin *Q. J. Microsc. Sci.* **89** 239–52

[17] Wolff H K 1952 Sur une methode de culture d'organes embryonnaires *in vitro Texas Rep. Biol. Med.* **10** 463–72

[18] Trowell O A 1954 A modified technique for organ culture *in vitro Exp Cell Res.* **6** 246–8

[19] Merchant D J, Kahn R H and Murphy W H 1964 *Handbook of Cell and Organ Culture* 2nd edn (Minneapolis, MN: Burgess)

[20] Chen J M 1954 The effect of insulin on embryonic limb-bones cultivated *in vitro J. Physiol.* **125** 148

Chlopin N G 1922 Ueber *in vitro* Kulturen der embryonalen Gewebe der Säugetiere *Arch. Mikroskop. Anat. Entwicklungsmech.* **96** 435

[21] Shaffer B M 1956 The culture of organs from the embryonic chick on cellulose acetate fabric *Exp Cell Res.* **11** 244–8

[22] Gantenbein B *et al* 2015 Organ culture bioreactors—platforms to study human intervertebral disc degeneration and regenerative therapy *Curr. Stem Cell Res. Ther.* **10** 339–52

[23] Barabanov V M 1990 The use of rafts made of polymers with hydrophobic properties in organ culture—a new modification of Chen's raft method *Ontogenez* **21** 315–8

[24] Willmer E N 2015 *Cells and Tissues in Culture: Methods, Biology and Physiology* (Amsterdam: Elsevier) pp 70–100
[25] Jones H H and Keeler D 1971 Organ culture method using a modified Grobstein raft technique *Clin. Orthop. Relat. Res.* **74** 273–8
[26] Spratt N T 1947 Development *in vitro* of the early chick blastoderm explanted on yolk and albumen extract saline–agar substrata *J. Exp. Zool.* **106** 345–65
[27] Khan F A 2014 *Biotechnology in Medical Sciences* (Boca Raton, FL: CRC Press)
[28] Wong C, Inman E, Spaethe R and Helgerson S 2003 Fibrin-based biomaterials to deliver human growth factors *Thromb. Haemost.* **89** 573
[29] Mosesson M W 2005 Fibrinogen and fibrin structure and functions *J. Thromb. Haemost.* **3** 1894
[30] Janmey P A, Winer J P and Weisel J W 2009 Fibrin gels and their clinical and bioengineering applications *J. R. Soc. Interface* **6** 1
[31] Anitua E, Prado R and Orive G 2013 Endogenous morphogens and fibrin bioscaffolds for stem cell therapeutics *Trends Biotechnol.* **31** 364
[32] Leighton J 1991 Radial histophysiologic gradient culture chamber rationale and preparation *In Vitro Cell. Dev. Biol.* A **27** 786–90
[33] Leighton J, Mark R and Rush G 1968 Patterns of three-dimensional growth in collagen coated cellulose sponge: carcinomas and embryonic tissues *Cancer Res.* **28** 286–96
[34] Sorour O, Raafat M, El-Bolkainy N and Mohamad R 1975 Infiltrative potentiality of brain tumors in organ culture *J. Neurosurg.* **43** 742–7
[35] Gullino P M and Knazek R A 1979 Tissue culture on artificial capillaries *Cell Culture Methods in Enzymology* ed W B Jakoby and I Pastan vol 58 (New York: Academic) pp 178–84
[36] Linser P and Moscona A A 1980 Induction of glutamine synthetase in embryonic neural retina—localization in Muller fibers and dependence on cell interaction *Proc. Natl Acad. Sci. USA* **76** 6476–81
[37] Twentyman P R 1980 Response to chemotherapy of EMT6 spheroids as measured by growth delay and cell survival *Eur. J. Cancer* **42** 297–304
[38] Mareel M M, Bruynell E and Storme G 1980 Attachment of mouse fibrosarcoma cells to precultured fragments of embryonic chick heart *Virchows Arch.* B **34** 85–97
[39] Boyd M *et al* 2001 A gene therapy/targeted radiotherapy strategy for radiation cell kill by [131I]MIBG *J. Gene Med.* **3** 165–72
[40] Boyd M *et al* 2004 An efficient targeted radiotherapy/gene therapy strategy utilising human telomerase promoters and radioastatine and harnassing radiation-mediated bystander effects *J. Gene Med.* **6** 937–47
[41] Tsao Y D, Goodwin T J, Wolf D A and Spaulding G F 1992 Responses of gravity level variations on the NASA/JSC bioreactor system *Physiologist* **35** S49–50
[42] Schwarz R P, Goodwin T J and Wolf D A 1992 Cell culture for three-dimensional modeling in rotating-wall vessels: an application of simulated microgravity *J. Tiss. Cult. Meth.* **14** 51–7
[43] Jessup J M, Goodwin T J and Spaulding G F 1993 Prospects for use of microgravity-based bioreactors to study three-dimensional host–tumor interactions in human neoplasia *J. Cell. Biochem.* **51** 290–300
[44] Goodwin T J, Prewett T L, Wolf D A and Spaulding G F 1993 Reduced shear stress: a major component in the ability of mammalian tissues to form three-dimensional assemblies in simulated microgravity *J. Cell Biochem.* **51** 301–11

[45] Bartis Domokos P J 2011 *Three Dimensional Tissue Cultures and Tissue Engineering* (Hungary: University of Pécs)

[46] He S Y, Lin Y H, Hou K Y and Hwang S C J 2011 Degradation of dimethyl-sulfoxide-containing waste water using airlift bioreactor by polyvinyl-alcohol-immobilized cell beads *Bioresour. Technol.* **102** 5609–16

[47] Yan S, Wang P, Zhai Z and Yao J 2011 Fuel ethanol production from concentrated food waste hydrolysates in immobilized cell reactors by *Saccharomyces cerevisiae* H058 *J. Chem. Technol. Biotechnol.* **86** 731–8

[48] Seol E *et al* 2011 Sustained hydrogen production from formate using immobilized recombinant *Escherichia coli* SH5 *Int. J. Hydrog. Energy* **36** 8681–6

[49] Panesar R, Panesar P S, Singh R S and Kennedy F J 2011 Hydrolysis of milk lactose in a packed bed reactor system using immobilized yeast cells *J. Chem. Technol. Biotechnol.* **86** 42–6

[50] Margaritis A, Merchant F J A and Abbott B J 1984 Advances in ethanol production using immobilized cell systems *Crit. Rev. Biotechnol.* **1** 339–93

[51] Manju N J, Deepesh V, Achuthan C, Rosamma P and Singh I S B 2009 Immobilization of nitrifying bacterial consortia on wood particles for bioaugmenting nitrification in shrimp culture systems *Aquaculture* **294** 65–75

[52] Chu C Y, Wu S Y, Hsieh P C and Lin C Y 2011 Biohydrogen production from immobilized cells and suspended sludge systems with condensed molasses fermentation solubles *Int. J. Hydrog. Energy* **36** 14078–85

[53] Genisheva Z, Mussatto S I, Oliveira J M and Teixeira J A 2011 Evaluating the potential of wine-making residues and corn cobs as support materials for cell immobilization for ethanol production *Ind. Crops Prod.* **34** 979–85

[54] Smidsrød O and Skjåk-Braek G 1990 Alginate as immobilization matrix for cells *Trends Biotechnol.* **8** 71–8

[55] Zimmermann H *et al* 2003 Fabrication of homogeneously cross-linked, function alalginate microcapsules validated by NMR-, CLSM- and AFM imaging *Biomaterials* **24** 2083–96

[56] Lemare F, Steimberg N, Le Griel C, Demignot S and Adolphe M 1998 Dedifferentiated chondrocytes cultured in alginate beads: restoration of the differentiated phenotype and of the metabolic response to interleukin-1β *J. Cell Physiol.* **176** 303–13

[57] Krol S, Nolte M, Diaspro A, Mazza D, Magrassi R, Gliozzi A and Fery A 2005 Encapsulated living cells on microstructured surfaces *Langmuir* **21** 705–9

[58] Tope A M, Srinivas N, Kulkarni S J and Jamil K 2001 Mesoporous molecular sieve (MCM-41) as support material for microbial cell immobilization and transformation of 2,4,6-trinitrotoluene (TNT): a novel system for whole cell immobilization *J. Mol. Catal.* B **16** 17–26

[59] Banerjee A and Ghoshal A K 2011 Phenol degradation performance by isolated *Bacillus cereus* immobilized in alginate *Int. Biodeterior. Biodegradation* **65** 1052–60

[60] Akdogan H A and Pazarlioglu N K 2011 Fluorene biodegradation by *P. osteratus*–Part II: Biodegradation by immobilized cells in a recycled packed bed reactor *Process Biochem.* **46** 840–6

[61] Yiu H H P and Wright P A 2005 Enzymes supported on ordered mesoporous solids: a special case of an inorganic–organic hybrid *J. Mater. Chem.* **15** 3690–700

[62] Corma A and Navarro M T 2002 From micro to mesoporous molecular sieves: adapting composition and structure for catalysis *Stud. Surf. Sci. Catal.* **142** 487–501

[63] Fan J *et al* 2005 Mesoporous silica nanoreactors for highly efficient proteolysis *Chem. Eur. J.* **11** 5391–6

[64] Wan Y and Zhao D Y 2007 On the controllable soft-templating approach to mesoporous silicates *Chem. Rev.* **107** 2821–60

[65] Takahashi H, Li B and Sasaki T 2000 Catalytic activity in organic solvents and stability of immobilized enzymes depend on the pore size and surface characteristics of mesoporous silica *Chem. Mater.* **12** 3301–5

[66] Niu X *et al* 2013 'Fish-in-net', a novel method for cell immobilization of *Zymomonas mobilis PLoS One* **8** e79569

[67] Hasegawa T, Sakamoto A, Wada A, Fukai T, Iida H and Ikeda S 2015 Keratinocyte progenitor cells reside in human subcutaneous adipose tissue *PLoS One* **10** e0118402

[68] Halleux C and Schneider Y J 1994 Iron absorption by Caco-2 cells cultivated in serum-free medium as *in vitro* model of the human intestinal epithelial barrier *J. Cell. Physiol.* **158** 17–28

[69] Limat A, Breitkreutz D, Thiekotter G, Klein E C, Braathen L R, Hunziker T and Fusenig N E 1995 Formation of a regular neo-epidermis by cultured human outer root sheath cells grafted on nude mice *Transplantation* **59** 1032–8

[70] Kondo S, Kooshesh F and Sauder D N 1997 Penetration of keratinocyte-derived cytokines into basement membrane *J. Cell Physiol.* **171** 190–5

[71] Maas-Szabowski N, Stark H J and Fusenig N E 2000 Keratinocyte growth regulation in defined organotypic cultures through IL-1-induced KGF expression in resting fibroblasts *J. Invest. Dermatol.* **114** 1075–84

[72] Maas-Szabowski N, Stark H J and Fusenig N E 2002 Cell interaction and epithelial differentiation *Culture of Epithelial Cells* 2nd edn ed R I Freshney and M G Freshney (Hoboken, NJ: Wiley) pp 31–63

[73] Chambard M, Vemer B, Gabrion J, Mauchamp J, Bugeia J C, Pelassy C and Mercier B 1983 Polarization of thyroid cells in culture: evidence for the basolateral localization of the iodide 'pump' and of the thyroid-stimulating hormone receptor–adenylcyclase complex *J. Cell Biol.* **96** 1172–7

[74] McCall E, Povey J and Dumonde D C 1981 The culture of vascular endothelial cells on microporous membranes *Thromb. Res.* **24** 417–31

[75] Elvin P, Wong V and Evans C W 1985 A study of the adhesive, locomotory and invasive behaviour of Walker 256 carcinosarcoma cells *Exp. Cell Biol.* **53** 9–18

[76] Repesh L A 1989 A new *in vitro* assay for quantitating tumor cell invasion *Invas. Metast.* **9** 192–208

[77] Schlechte W, Brattain M and Boyd D 1990 Invasion of extracellular matrix by cultured colon cancer cells: dependence on urokinase receptor display *Cancer Commun.* **2** 173–9

[78] Brunton V, Ozanne B, Paraskeva C and Frame M 1997 A role for epidermal growth factor receptor, c-Src and focal adhesion kinase in an *in vitro* model for the progression of colon cancer *Oncogene* **14** 283–93

[79] Seeds N W 1971 Biochemical differentiation in reaggregating brain cell culture *Proc. Natl Acad. Sci. USA* **68** 1858–61

[80] Trapp B D, Honegger P, Richelson E and Webster H 1981 Morphological differentiation of mechanically dissociated fetal rat brain in aggregating cell cultures *Brain Res.* **160** 235–52

[81] Bjerkvig R, Laerum O D and Mella O 1986 Glioma cell interactions with fetal rat brain aggregates *in vitro*, and with brain tissue *in vivo Cancer Res.* **46** 4071–9

[82] Jackson J P, Tonge P D and Andrews P W 2007 Techniques for neural differentiation of human EC and ES cells *Culture of Human Stem Cells* ed R I Freshney, G N Stacey and J M Auerbach (Hoboken, NJ: Wiley) pp 61–91
[83] Bjerkvig R, Steinsvag S K and Laerum O D 1986 Reaggregation of fetal rat brain cells in a stationary culture system. I: Methodology and cell identification *In Vitro* **22** 180–92
[84] Grinnell F 2008 Fibroblast mechanics in three-dimensional collagen matrices *J. Bodyw. Mov. Ther.* **12** 191–3
[85] Bissell M J, Hall H G and Parry G 1982 How does the extracellular matrix direct gene expression? *J. Theor. Biol.* **99** 31–68
[86] Yang J, Balakrishnan A, Hamamoto S, Beattie C W, Das Gupta T K, Wellings S R and Nandi S 1987 Different mitogenic and phenotypic responses of human breast epithelial cells grown in two versus three dimensions *Exp. Cell Res.* **167** 563–9
[87] Lin C Q and Bissell M J 1993 Multi-faceted regulation of cell differentiation by extracellular matrix *FASEB J.* **7** 737–43
[88] Smalley K S, Lioni M and Herlyn M 2006 Life isn't flat: taking cancer biology to the next dimension *In Vitro Cell. Dev. Biol. Anim.* **42** 242–7
[89] Grinnell F 1976 Biochemical analysis of cell adhesion to a substratum and its possible relevance to cell metastasis *Prog. Clin. Biol. Res.* **9** 227–36
[90] Mazzoleni G, Di Lorenzo D and Steimberg N 2009 Modeling tissues in 3D: the next future of pharmaco-toxicology and food research? *Genes Nutr.* **4** 13–22
[91] Horning J L, Sahoo S K, Vijayaraghavalu S, Dimitrijevic S, Vasir J K, Jain T K, Panda A K and Labhasetwar V 2008 3-D tumor model for *in vitro* evaluation of anticancer drugs *Mol Pharm.* **5** 849–62
[92] Edmondson R, Broglie J J, Adcock A F and Yang L 2014 Three-dimensional cell culture systems and their applications in drug discovery and cell-based biosensors Assay *Drug Dev. Technol.* **12** 207–18
[93] Langhans S A 2018 Three-Dimensional *in vitro* cell culture models in drug discovery and drug repositioning *Front. Pharmacol.* **9** 6
[94] O'Leary R, Arrowsmith M and Wood E J 2002 Characterization of the living skin equivalent as a model of cutaneous re-epithelialization *Cell Biochem. Funct.* **20** 129–41
[95] Halim A S, Khoo T L and Mohd. Yussof S J 2010 Biologic and synthetic skin substitutes: an overview *Indian J. Plast. Surg.* **43** S23–8
[96] Vig K, Chaudhari A, Tripathi S, Dixit S, Sahu R, Pillai S, Dennis V A and Singh S R 2017 Advances in skin regeneration using tissue engineering *Int. J. Mol. Sci.* **18** 789
[97] Metcalfe A D and Ferguson M W 2007 Tissue engineering of replacement skin: the crossroads of biomaterials, wound healing, embryonic development, stem cells and regeneration *J. R. Soc. Interface.* **4** 413–37
[98] Yang E K, Seo Y K, Youn H H, Lee D H, Park S N and Park J K 2000 Tissue engineered artificial skin composed of dermis and epidermis *Artif. Organs* **24** 7–17
[99] Dixit S *et al* 2017 Immunological challenges associated with artificial skin grafts: available solutions and stem cells in future design of synthetic skin *J. Biol. Eng.* **11** 49
[100] Haraguchi Y, Kagawa Y, Sakaguchi K, Matsuura K, Shimizu T and Okano T 2017 Thicker three-dimensional tissue from a 'symbiotic recycling system' combining mammalian cells and algae *Sci. Rep.* **7** 41594
[101] Bhatia S, Goli D, Naved T and Sharma A 2018 Nutraceutical properties of Indian seaweed *Porphyra Adv. Invest. Pharmacol. Ther. Med.* **1** 47–54

[102] Bhatia S 2016 *Natural Polymer Drug Delivery Systems: Nanoparticles, Plants, and Algae* (Cham: Springer) pp 117–27

[103] Bhatia S and Goli D 2016 *Leishmaniasis: Biology, Control and New Approaches for Its Treatment* (Boca Raton, FL: CRC Press) pp 164–73

[104] Bhatia S, Namdeo A G and Nanda S 2010 Factors effecting the gelling and emulsifying properties of a natural polymer *Sys. Rev. Pharm.* **1** 86–92

[105] Bhatia S 2018 *Introduction to Pharmaceutical Biotechnology: Enzymes, Proteins and Bioinformatics* vol 2 (Bristol: IOP Publishing) pp 172–7

[106] Bhatia S, Garg A, Sharma K, Kumar S, Sharma A and Purohit A P 2011 Mycosporine and mycosporine-like amino acids: a paramount tool against ultra-violet irradiation *Pharmacogn. Rev.* **5** 138–46

[107] Bhatia S, Rathee P, Sharma K, Chaugule B B, Kar N and Bera T 2013 Immuno-modulation effect of sulphated polysaccharide (porphyran) from *Porphyra vietnamensis Int. J. Biol. Macromol.* **57** 50–6

[108] Bhatia S *et al* 2008 Novel algal polysaccharides from marine source: *Porphyran Pharmacogn. Rev.* **2** 271–6

[109] Bhatia S, Sharma K, Namdeo A G, Chaugule B B, Kavale M and Nanda S 2010 Broad-spectrum sun-protective action of *Porphyra*-334 derived from *Porphyra vietnamensis Pharmacogn. Res.* **2** 45–9

[110] Bhatia S 2016 *Introduction to Pharmaceutical Biotechnology: Basic Techniques and Concepts* vol 1 (Bristol: IOP Publishing) pp 167–74

[111] Bhatia S *et al* 2014 Significance of algal polymer in designing amphotericin B nanoparticles *Sci. World J.* **2014** 564573

[112] Bhatia S *et al* 2015 Investigation of the factors influencing the molecular weight of Porphyran and its associated antifungal activity *Bioact. Carbohydr. Diet. Fiber* **5** 153–68

[113] Bhatia S, Sharma K, Sharma A, Nagpal K and Bera T 2015 Anti-inflammatory, analgesic and antiulcer properties of *Porphyra vietnamensis Avicenna J. Phytomed.* **5** 69–77

[114] Bhatia S, Sharma K and Bera T 2015 Structural characterization and pharmaceutical properties of porphyran *Asian J. Pharm.* **9** 93–101

[115] Bhatia S and Bera T 2015 Evaluation of pharmacognostical, phytochemical and anti-microbial properties of *Porphyra vietnamensis Int. J. Green Pharm.* **9** 131–7

[116] Bhatia S, Sharma K, Sharma A, Namdeo A G and Chaugule B B 2011 Anti-oxidant potential of Indian *Porphyra Pharmacology* **1** 248–57

[117] Bhatia S *et al* 2015 *Modern Applications of Plant Biotechnology in Pharmaceutical Sciences* (New York: Academic, Elsevier) pp 164–74

[118] Bhatia S 2016 *Systems for Drug Delivery: Safety, Animal, and Microbial Polysaccharides* (Switzerland: Springer) pp 122–7

[119] Bhatia S 2016 *Nanotechnology in Drug Delivery: Fundamentals, Design, and Applications* (Boca Raton, FL: CRC Press) pp 121–7

[120] Shinohara K *et al* 1988 Algal phycocyanins promote growth of human cells in culture *In Vitro Cell. Dev. Biol.* **24** 1057–60

[121] Bhatia S 2018 *Introduction to Pharmaceutical Biotechnology: Enzymes, Proteins and Bioinformatics* vol 2 (Bristol: IOP Publishing)

[122] Bhatia S 2018 *Introduction to Pharmaceutical Biotechnology: Basic Techniques and Concepts* vol 1 (Bristol: IOP Publishing)

IOP Publishing

Introduction to Pharmaceutical Biotechnology, Volume 3
Animal tissue culture and biopharmaceuticals
Saurabh Bhatia, Tanveer Naved and Satish Sardana

Chapter 3

Stem cell culture

3.1 Introduction

Stem cells are the cells that retain their proliferative potential, always balancing between self-renewal and differentiation throughout their lives. Most cells, when removed from the body and maintained in culture, usually maintain their original character. However, when stem cells are cultured, as well as when they are present in tissues, they may divide continuously or they can develop into one or more cell types. Every stem cell always functions for the renewal of one specific tissue type. Previously, regeneration via stem cells has been considered more challenging in more sensitive and complicated tissue such as the brain [1]. Recently, a little hope has apparently been generated by advancements in animal tissue culture science which allow the replacement of injured nerve cells in the human brain through developing new ones. Under *in vivo* conditions stem cells are present in 3D microenvironments called stem cell niches. These 3D environments can be developed under *in vitro* conditions using 3D biomaterial scaffolds which exactly imitate these microenvironments, offering advantages over earlier 2D culture techniques, such as using polystyrene in the regeneration and replacement of tissues [2]. The polystyrene mediated 2D tissue culture technique has now been replaced by 3D biomaterial scaffolds, because the latter allows more precise development of cell polarity in the microenvironment and the cells exhibit biochemical and mechanical properties comparable to soft tissue [3]. Several 3D environments have been developed using naturally derived and synthetic biomaterial scaffolds to encourage stem cell growth. The 3D synthetic scaffolds offer greater reproducibility. Natural biomaterials are usually made up of proteins and polysaccharides and can be utilized to mimic and eventually replace the extracellular matrix. This matrix contains binding sites for cell adhesion and encourages cell growth. Recently, fibrin scaffolds have been developed for several tissue engineering applications [4]. This type of scaffold can be developed by polymerizing the protein fibrinogen derived from blood plasma, eventually to utilize in drug delivery systems to further control the release of therapeutic factors

doi:10.1088/2053-2563/aafac0ch3

[5, 6]. Thus stem cells can potentially be utilized to regenerate tissue. There are different types of stem cells, as discussed below.

3.2 Embryonic stem cells

The inner cellular mass of an embryo is made up of embryonic stem (ES) cells. These cells have the potential to divide indefinitely in culture and can differentiate into any cell type. This feature can be utilized to maintain cell lines from ES cells. For research purposes, ES cells derived from mouse blastocysts are often utilized. In the future, perhaps human ES (hES) cells might serve as potential source of cells for transplantation treatments to replace degenerated or malfunctioning cells [7]. Moreover, genetically engineered embryonic stem cells can also be utilized as vectors to carry and express genes in targeted tissue or organs following transplantation or regeneration in the course of gene therapy [7].

Since ES cells are different to adult cells, characterization by suitable methods is required to ensure their cell type. Characterization ensures that adult cells derived from ES cells have similar characteristics to the parent/adult cells. One of the main objectives of hES cell research is the replacement of injured tissue, in particular in the case of degenerative diseases. However, the underlying signaling mechanisms involved in the lineage restriction of hES cells to accept various cellular phenotypes are still under study. For the further development of hES cell-based treatments, in particular towards clinical applications, suitable culture environments should be developed to create genetically stable homogeneous populations of cells. This development allows the production of homogenous cells which can further obstruct possible harmful effects following transplantation [8].

These embryogenesis mimicking cells can develop into cells of the three embryonic germ layers: endoderm, mesoderm and ectoderm. These three layers develop into various specialized cell types of the entire organism [9–12]. As discussed above, hES cells can develop into all terminal cell types of the human body. Thus hES can be considered as an ideal model for screening small molecules in the form of drugs, peptides or proteins. This makes them potentially valuable for pharmaceutical development and safety assessment, particularly for discovering new chemical molecules that modulate the destiny of adult stem cells [9–12].

There is variation in the degree of plasticity or potency of stem cells, as well as their potential for long-term proliferation. Pluripotent stem cells, either derived from *in vitro* fertilized eggs, such as ES cells, or reprogrammed from somatic cells, such as induced pluripotent stem cells (iPSCs), can divide continuously and can develop into to all cell types of the adult body. iPSCs are somatic stem cells which are artificially reprogrammed in the laboratory with transcription factors to express the pluripotent properties of embryonic stem cells.

Mouse embryonic stem cells have been investigated for many years, in particular in drug discovery, to develop a model organism in the form of genetically modified mice. Thus, mouse embryonic stem cells can be utilized for the following purposes:

- *Target validation or target identification.* Stem cells can be considered as faster tools for the discovery of novel targets. Predominantly used in target validation and identification, comparing patient-derived iPSCs to normal cells can offer scientists enough understanding to determine physiological deviations at the cellular level [10–12]. Since ES cells and iPSCs display inherited traits and disorders, scientists can utilize them to better understand the basic mechanisms of diseases and make disease models more descriptive of the actual disease than animal models [10–12].
- *Target selectivity.* For more promising treatment by the new generation of therapeutics, selective targeting of cancer stem cells is required. However, examinations of both human cancer and normal stem cells are dependent on the utilization of advanced robust technology, which is limited. Using drug discovery tools it is now possible to tell the difference between cancerous and normal human pluripotent stem cells. To achieve this researchers have identified small molecules from databases of known compounds that encourage differentiation to overcome neoplastic self-renewal. Unexpectedly, an antipsychotic drug (thioridazine) selectively targets the neoplastic cells, and damages human somatic cancer stem cells, those that are capable of *in vivo* leukemic disease initiation, while having no effect on normal stem cells [13–19].
- *Pluripotent stem cells and iPSCs.* Human embryonic stem cells have the capability to become any cell in the body, which means they have multiple potentials, i.e. they are pluripotent. There are four types of pluripotent human stem cells: embryonic stem cells, nuclear transplant stem cells, parthenote stem cells and induced stem cells. In 2007 scientists first developed pluripotent stem cells. Research has revealed that of 20 000 protein encoding genes present in stem cells, only a few are responsible for pluripotency and self-renewal. Although pluripotent genes are also present in adult cells, they are turned off, and we do not yet known how to turn them on. Recently, several researchers decided to introduce an extra copy of a synthesized pluripotent gene in normal adult cells by infecting them with a virus. Once infected, some cells will become iPSCs with characteristics similar to embryonic stem cells (figure 3.1).

  These cells have proven that the introduction of four transcription factors into fibroblasts reverts them to a pluripotent state. However, when reprogramming, the epigenetic state of the differentiated cell is only somewhat rearranged, which eventually disturbs the differentiation potential of subsequent cell lines. If this has significant downstream functional effects in *in vitro* assays in which the cells are later employed remains to be determined. Further understanding of both ES cells and iPSCs is required to determine the consequences of epigenetic variation and to better understand the differences between the cells.
- *Toxicity evaluation* [19–24]. The appearance of safety concerns late in the drug development process is considered as one of the major reasons for drug attrition. The current screening tools established to determine adverse effects

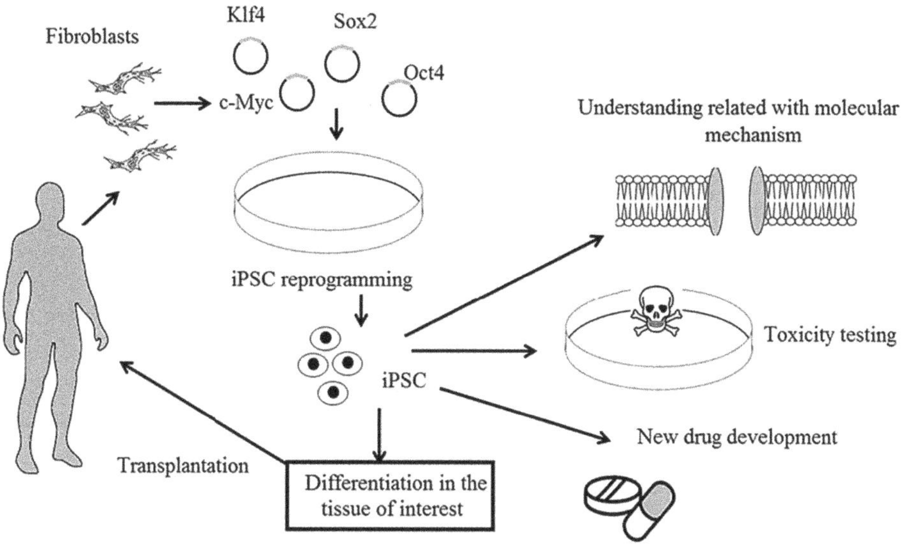

Figure 3.1. Reprogramming of normal fibroblasts into induced pluripotent stem cells.

often require previous understanding of the target's tendency for such effects, drawn from clinical data or animal studies. Traditionally, toxicity investigations are usually performed on immortalized cell lines expressing targets of interest. Stem cells offer numerous advantages, such as being of human origin and thus permitting various investigations of molecules such as proteins, peptides and drugs and their interactions in a more appropriate physiological setting. Moreover, genetic manipulation in stem cells has allowed the development of models such as fluorescent reporters, tagged proteins which are suited for specific assays such as high content analysis and high content screening. Toxicity examination can be performed by means of generic human pluripotent stem cell lines to generate various cell types of interest, or it can be expanded to include several lines with varying and selected genetic backgrounds. In association with hES cell technology, current drug discovery may shortly have a physiologically relevant screening tool that displays normal growth and genetic structure.

The stem cell fate, its self-renewal or differentiation potential, can be determined by exploring molecular mechanisms. This will considerably support our insight into the therapeutic potential of stem cells [25–29]. Until now, most researchers focused on signal transduction and the associated events at the molecular level throughout ES cell development. Reports have demonstrated that the niche of a stem cell, its microenvironment, is active, not static, and can change or even be created whenever required [30–34]. Consequently, stem cell function is controlled by small molecules that can govern the cell's destiny, which provides an appropriate microenvironment to screen drugs.

- *Drug screening*. The use of stem cells and derived models thereof for drug discovery relies on efficient and faithful *in vitro* reproduction of disease phenotypes. Human pluripotent stem cells have been used to recreate many disease phenotypes. The development of new drugs is very expensive, cumbersome and tedious. In particular, the early developmental phases of research require preliminary trials or pilot studies, assessing drugs through *in vitro* models, ultimately screening millions of compounds based on their therapeutic activity and toxicity. Therefore, an appropriate model is required to determine the effects and safety of molecules entered in the process. Preclinical investigation, in particular cell-based *in vitro* examinations with high human relevance, are required to serve as a screening platform to identify low-molecular-weight compounds that affect endogenous stem cell populations and repair damaged tissue [10, 35–37]. Moreover, numerous cell lines, e.g. H1, H7, H9, D3, mES and hES, cells are considered as ideal models for drug screening [10, 35–37].

3.3 Epithelial stem cells

Owing to the presence of multipotent stem cells and/or unipotent progenitor cells, most epithelial tissues have a self-renewal capacity which allows them to replace older cells. Epithelial cells, which are present over the outer surfaces of the body, continuously die and are shed. This damage is always compensated for by a continuous replacement process, referred to as tissue homeostasis, and is critical for the maintenance of adult tissues. Usually, this process of homeostasis is controlled by stem cells. Tissue homeostasis occurs in a highly organized and regular fashion. The self-renewal potential of stem cells allows cells to develop into the cell lines of their tissue of origin [38].

After activation, epithelial stem cells can produce proliferating progeny, referred to as transiently amplifying cells. Under normal physiological conditions transiently amplifying cells will proliferate vigorously for a limited period of time, to further increase the cellular pool that will then develop along a particular cell line to form the tissue. The process of shedding epithelial cells and their replacement is a continuous process which has been observed in mammals, in particular in the mouse, in which replacement usually takes 3–4 days. During development these epithelial stem cells remain specified and controlled by epithelial–mesenchymal interactions resulting in effective communications between the epithelium and the mesenchyme with its heterotypic cell population. Despite the functional and morphological variations between the epithelial tissues, they share common signaling pathways. A deeper understanding of these pathways helps in the *in vitro* development of epithelial stem cells, which may further help in controlling epithelial stem cell maintenance, activation, lineage determination and differentiation. Moreover, deregulation of these pathways can lead to human disorders, including cancer. Understanding epithelial stem cell biology has major clinical implications for the diagnosis, prevention and treatment of human diseases, as well as for regenerative medicine.

In cellular biology, in particular developmental biology and stem cell biology, the isolation and long-term maintenance of the primary cells, mainly stem/progenitor populations, is an essential step that involves various fundamental approaches. One approach is the maintenance and development of stratified and columnar epithelial cells that are highly regenerative, and disproportionately account for several human cancers. However, adult stem cell cloning and utilization is limited due to the challenges in maintaining these cells under *in vitro* environments, in particular in an immature state. Recent scientific breakthroughs have resulted in rapid developments in stem cell biology. These advancements allowed the use of small molecules and growth factors to offer a 3D *in vivo* environment that mimics tissue niche environments and allows an 'organoid culture' [39]. The first culture of human adult stem cells was established in 1975 using human keratinocytes [40, 41]. In this work, Rheinwald and Green developed human keratinocyte cultures and maintained them for prolonged periods in the presence of a sublethally irradiated mouse fibroblast cell line (3T3-J2). This cell line is a subclone of the original mouse embryonic fibroblast line (3T3-Swiss Albino). After irradiation of these cells, they are usually used as a feeder layer for the growth of human keratinocytes. During this investigation, Rheinwald and Green did not termed the cloned keratinocytes grown on 3T3 cell as 'stem cells'. However, after their observations related to the remarkable capacity of these cells to divide and form new colonies after each passage, they named them 'holoclones' [42]. These small and immature cells showed intense nuclear staining with p63, a master regulator of stratified epithelial cells, which is often suppressed in order for these cells to differentiate [43]. Stratified tissues such as skin, lung bronchia, mammary glands and bladder urothelium contain stem cell population predominantly localized in the basal layer. These cells were stained with p63, consistent with *in vitro* studies [44].

The tumor suppressor p63 seems quite similar to p53 in function and exhibits a high extent of evolutionary conservation. It has been reported that p63 is found in the stem cells of the proliferative compartment, but not in the transit amplifying keratinocytes that have exited the compartment [45–49].

Autologous skin grafts are effective for the repair and regeneration of large skin wounds, however, the accessibility of large amounts of skin is often restricted. In this context human keratinocytes can be utilized for skin grafting to successfully graft burnt skin and regenerate permanent epidermis that results from split-thickness skin grafts, a skin graft including the epidermis and part of the dermis [6, 7].

Several bioengineering procedures including several autologous skin substitutes have been engineered for human clinical utilization but only a few skin substitutes are available for use in animals. These advanced bioengineering procedures involve keratinocytes and fibroblast cells in the form of biosynthetic skin cultured on biomaterial scaffolds and can be effectively utilized to treat wounds with delayed healing [50]. For repairing large and deep skin wounds, using the patient's own cells is considered the best approach for reconstructing skin [51]. Under an *in vitro* environment using certain advanced culture-based techniques it is possible to amplify the number of cells which can be utilized later to treat large skin lesions. To date, most skin-based clinical studies are conducted on animals, where they are considered as a model for wound healing in human beings. Several advanced procedures for the development of skin by artificial means, in particular reconstruction of human skin, involves grafts onto athymic mice, which has proved to be an effective approach for the permanent dressing of burns [52, 53]. Similarly, spraying cultured autologous keratinocytes onto a skin defect is an effective approach and has shown wonderful results in a pig model with improved epidermal thickness, confluence and blood vessel formation in the graft [54].

A similar procedure has been applied for human corneal epithelial cell development and application for transplantation [45, 55, 56]. Green and co-workers, for the first time, established and tested a procedure for cloning human adult stem cells in the fields of basic biology and regenerative medicine. Novel culture approaches for the development of stratified epithelial cells and columnar epithelial cells under *in vitro* environments can be utilized for regeneration and to treat human diseases, or can be considered as a model *in vitro*.

3.3.1 Stratified epithelial cell culture

Self-renewing p63+ cells, present in the basement membrane, in particular in stratified epithelial tissues, including glandular and pseudostratified epithelium, can self-renew to maintain stem/progenitor cells and develop into progeny that form functional tissues [41, 57]. As mentioned above, irradiated fibroblasts can be used as a feeder layer in co-culture systems, such as that of Rheinwald and Green (1975), who developed an *in vitro* model for keratinocyte cell cultures in which the use of murine fibroblasts as a feeder layer was introduced. These cells are modified fibroblasts, the presence of which allows the keratinocyte cells to remain proliferative for longer periods of time [41, 57].

This discovery of co-culturing cells with irradiated mouse 3T3-J2 fibroblasts has led to the cloning and expansion of epithelial stem cells, such as skin keratinocytes and corneal epithelial cells. However, this standard procedure is still limited to the long-term culture of keratinocytes and corneal cells [40, 41, 58].

A study related to the cloning of thymic epithelia stem cells was reported recently, where the thymic epithelial stem cells can be derived from diverse species, such as human cells when cultured with a 3T3 feeder system [43, 59, 60]. Additionally, researchers have recently used the 3T3 feeder procedure to isolate urothelial stem cells that expressed sonic hedgehog (the protein that in humans is encoded by the 'sonic hedgehog' (SHH) gene) and are present in the basal layer of the bladder urothelium [61]. Co-culture of these urothelial stem cells with a 3T3 feeder layer resulted in multiple cell lineages, including p63+ basal cells and Uroplakin 2+ and 3+ urothelial cells. Similarly, in 2011 researchers explored the 3T3 culture system to isolate three types of human airway epithelial stem cells, i.e. nasal, tracheal and distal airway stem cells. During the study they discovered that airway epithelial stem cells exhibit different cellular phenotypes, however, the immature stem cell clones seemed to be morphologically indistinguishable. Following the same process, it has been demonstrated that mouse tracheal transplantation and distal airway epithelial stem cells were readily incorporated into H1N1 influenza-damaged lung tissue and developed into different epithelial cells, while transplanted tracheal stem cells were restricted to only key airways [62] (figure 3.2). In an another study it was reported that human esophagus endoscopic biopsy samples derived clonogenic stem cells with the potential to differentiate into stratified squamous epithelia-like structures in an air–liquid interface culture system [63] (figure 3.2).

It has also been reported that Rho-associated protein kinase inhibitor seeded with 3T3 feeder cells considerably increased the proliferative capacity of epithelial stem cells. This process is called 'conditional reprogramming' [64, 65]. The potential to produce epithelial stem cell cultures from patients offers significant potential for cell-based diagnostics and therapeutics [66]. Currently, several researchers have demonstrated that the TGFβ/BMP/SMAD signaling pathway is vital in numerous epithelial tissues, such as ectoderm-derived skin and mammary gland tissue, endoderm-derived esophagus and prostate tissue, and mesoderm-derived epididymis. They revealed that the dual inhibition of pathways allowed the stable multiplication of human and mouse epithelial basal cell populations. Unexpectedly, this dual inhibition allowed the healthy growth of epithelial stem cells without the need for mouse 3T3 feeder cells.

Taken together, these scientific advancements for establishing co-culture systems, in particular in combination with small molecules and feeder cells, can be utilized to constantly and efficiently proliferate stratified epithelial stem/progenitor populations *in vitro*. An additional discovery in stratified epithelial culture, organoid culture, has been employed to proliferate and expand both basal and luminal human prostate progenitors. These human luminal precursors in the form of proliferators were multipotent and engineered in prostate gland-like structures *in vitro* [67]. However, producing 3D structures made up of stratified or pseudostratified epithelia to achieve reliable *in vivo* structural design remains an exciting challenge, while several

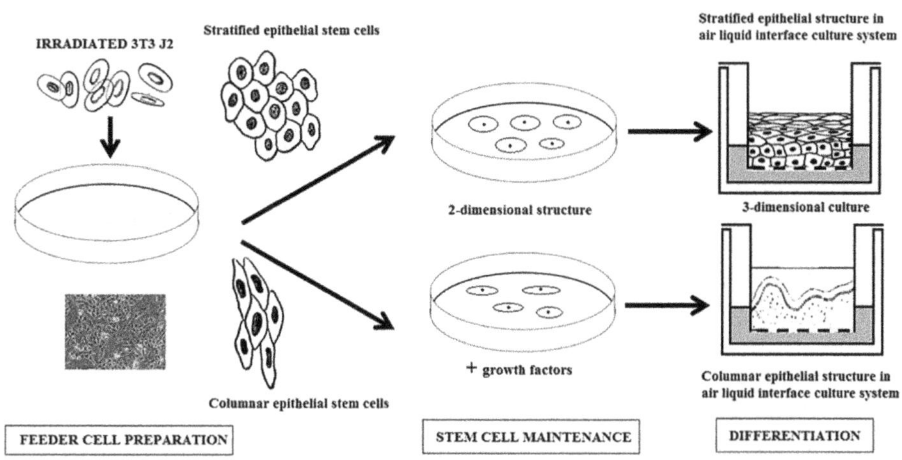

Figure 3.2. Co-culture system for human stratified and columnar epithelial stem cells on a 3T3 mouse feeder layer. Stratified epithelial stem cells derived from biopsy or surgical specimens are plated on a 3T3 layer for a long-term culture and columnar epithelial stem cells are plated on a 3T3 layer with defined factors which are essential for stem cell growth and maintenance.

scientists have demonstrated spheroid and organoid cultures. This issue can be resolved by establishing a procedure to allow self-organization, as performed in pluripotent stem cell-derived tissues [68, 69].

3.3.2 Columnar epithelial cell culture

Intestinal stem cells exhibit considerable potential to multiply at a high rate to maintain the *in vivo* intestinal epithelia niche. Similarly, hepatocytes exhibit high regenerative potential against any hepatic injury or damage, however, their capability to clone stem cell populations from columnar epithelial cells is limited, perhaps because of a lack of tissue niche signals *in vitro*. Recently, researchers have discovered an intestinal stem cell marker, named leucine-rich repeat-containing G-protein coupled receptor 5, in a sophisticated mouse model (Lgr5-EGFP-ires-CreERT2 mice crossed with the Cre-activated Rosa26 LacZ reporter). This mouse model, in the form of an intestinal organoid culture, was made up of villus-like structures and crypt-like zones with multiple intestinal cell types to mimic the *in vivo* environment [70]. In the presence of small molecules in the form of intestinal markers and growth factors, the stem cell fraction was suspended in Matrigel and cultured long-term [71]. Human epithelial cells derived from the small intestine and colon can be maintained infinitely under *in vitro* conditions by altering the *in vivo* niche using of nicotinamide, a p38 and TGFβ receptor inhibitor [72, 73]. This procedure can be utilized to grow and maintain other types of cells, such as pancreatic duct cells [74] and hepatocytes [75, 76], and has facilitated revolutionary advances in columnar epithelial cell culture.

3.4 Cancer cell culture

As discussed in chapter 1, in 1951 the first cancer cell line, the HeLa cell line, was derived from the cervical cancer patient Henrietta Lacks [77]. After this breakthrough cancer cell lines have been developed from different cancers and have been extensively utilized to investigate the pathobiology of cancer. They also offer the prospect to produce *in vivo* xenograft models and are used to examine anti-cancer drugs under *in vitro* and *in vivo* conditions. Even after remarkable advancements in cancer biology using cancer cell lines, the outcomes derived using these cells cannot satisfactorily display the complexity of the disease as originally expected, as cancer displays interpatient and intratumor heterogeneity. This was discovered by recent advances in next-generation sequencing [78]. Recently, researchers have engineered patient-derived xenograft models of breast cancer in nonobese diabetic severe combined immunodeficiency mice. This xenograft maintained the important characteristics of the original tumors and displayed metastatic capacity to specific sites [79]. This procedure can be adopted well to precisely determine cancer phenotypes, including the patient's gene mutation status and pathology. Discovery of the breast cancer model has now offered a way for the establishment of patient-derived xenograft models for different solid tumors. This can be projected to increase the preclinical examination of new cancer treatments and help in understanding the need for 'personalized medicine'.

3.5 Maintenance of stem cells

Stem cells should be maintained under *in vitro* conditions in such a way that they possess similar characteristics and differentiating abilities once they are transplanted into the tissue *in vivo*. Epidermal cells can be maintained by co-culturing with a 3T3 feeder layer, e.g. co-culturing of primary rat hepatocytes and murine Swiss 3T3 feeder cells. Cultures of growth-arrested feeder cells play a major role in the maintenance of stem cells by encouraging cell proliferation, mainly with low-density inocula. Fundamentally, feeder cells are defined as a layer of cells which are unable to divide, but stimulate other cells to proliferate by secreting extracellular secretions. This concept has been used to study the interactions between cell populations and is fundamental to cell–cell interaction studies of any kind [80].

Co-culturing of feeder cells always encourages growth of only one cell type by discharging growth factors in the culture medium. However, feeder cells also promote the growth of target cells through several other approaches. Prior to their addition into the main culture, the feeder layer cells are irradiated or treated with chemicals in order to inhibit their proliferation capacity. It has been observed that the addition of feeder cells is mandatory for some cell cultures, whereas in other cases the proliferation cells can be attained with just a conditioned medium. To understand this more clearly it is important to understand the development of feeder cells, as shown in figure 3.3. Different procedures have been developed to prevent feeder cell proliferation. Traditional treatments include mitomycin or γ-irradiation (figure 3.3). Currently, many other methods are also utilized, such as electric pulses or chemical fixation. These adherent growth-arrested feeder layer cells consist of viable and bioactive cells which form the substratum to condition the medium. Over this substratum desirable cells are allowed to culture, particularly at low clonal density. Often the cells of the feeder layer are irradiated or otherwise treated so that they will not proliferate. Feeder cells have the potential to encourage *in vitro* survival of certain fastidious cells (those that have complex nutritional requirements). Moreover, it also promotes growth of these cells, which requires a number of known or unknown soluble or membrane-bound growth factors and receptors. It has been reported that the growth and survival of certain cells is completely dependent on physical contact with a feeder layer, whereas certain cells can be cultured in a feeder-free environment. Culture dishes should be coated with extracellular matrix proteins such as fibronectin, laminin, collagen, or a blend of the extracellular matrix components called Matrigel.

By using feeder cells to encourage the growth of desirable cells, it is possible to achieve long-term maintenance of cells. Moreover, this approach is also helpful in retaining the proliferative and differentiating potential of desirable cells. In particular, for maintaining epidermal cells, supplementation of serum-free media with additional growth factors is required for the efficient growth of the cells. Similarly, feeder cells can also be utilized for the maintenance of non-epidermal epithelia cells, e.g. the epithelial cells of the prostate glands can be successfully cultured and maintained in the presence of a 3T3 feeder layer.

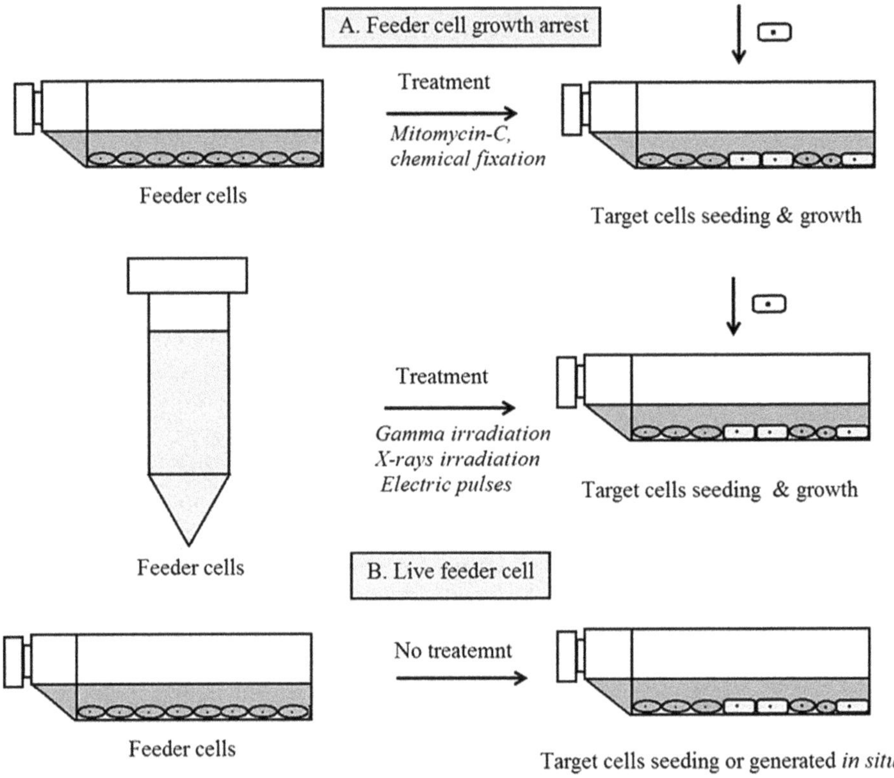

Figure 3.3. Treatments to prepare feeder cells.

3.6 Adult stem cells

Adult stem cells are inactive cells with a more limited self-renewal and differentiation capacity. Since each and every tissue has its own individual compartment of stem cells, adult tissues contain precursor cells which are responsible for replacing damaged cells within a given organ. Adult stem cells are capable of long-term self-renewal and have the differentiating potential to develop into mature cell types with specialized functions. This feature reveals their potential for asymmetric division, where the stem cell remains as a self-renewing stem cell, while the daughter cells are dedicated to replicating and differentiating into a mature cell type. In such cases, the cells produced are referred to as precursor or progenitor cells. After several rounds of mitosis, these precursor cells develop into mature cells, as shown in figure 3.4.

The stem cells' destiny is examined by their cellular interactions with their niche. The niche or microenvironment also includes other components such as stromal cells, extracellular matrix and signaling factors. Thus the interaction of a stem cell with these basic components and with other stem cells (inter-stem-cell communication) decides its properties and potential. As discussed above, adult stem cells proliferate asymmetrically to produce the next generation in the form of two daughter cells. Of these two cells one remains in the microenvironment as a self-

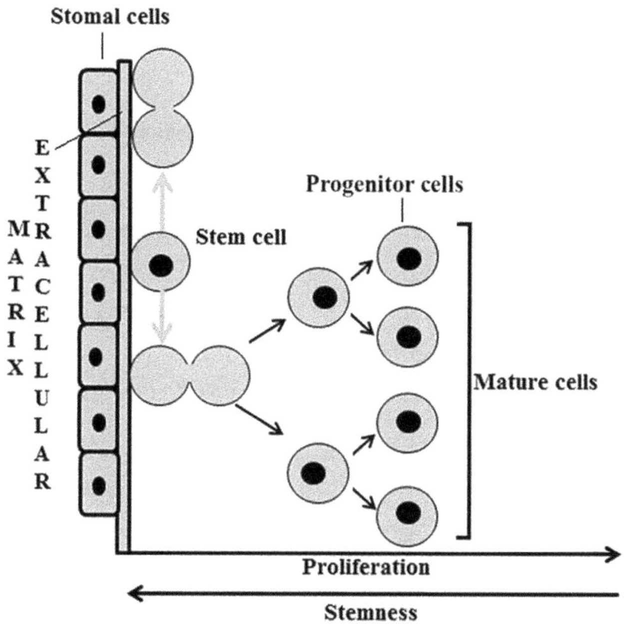

Figure 3.4. Stem cell microenvironment and its irregular cell division.

renewing stem cell, whereas the other becomes a precursor or progenitor cell, departs from the microenvironment, and moves into a pathway of proliferation and differentiation, eventually resulting in the development of a mature cell type.

3.7 Applications of cultured stem cells

The human system contains almost 200 different cell types that are present in the form of tissues and organs, to ultimately offer the functions required for growth and development (figure 3.5). Previously, most biologists were interested in the events that occur before birth. Later, the interest of some researchers moved towards developmental biology at the molecular level, i.e. the regulatory pathways that control specification and morphogenesis of tissues [81].

Significant evolution of stem cell research occurred when biologists' interest inclined towards knowing how tissues develop and are maintained in adult life, rather than how various cells arise in the embryo. This tremendous interest of researchers in the maintenance of adult tissues resulted in the exploration of its application against several diseases, in particular cancer. It became clear that some tissues have cellular heterogeneity, such as blood, skin and intestinal epithelium, and these tissues have differentiated cells with short life spans that are incapable of self-renewal. This has resulted in the conclusion that such types of tissues are maintained by stem cells. These stem cells have the capacity of renewal and the potential to produce daughter cells that experience further differentiation [82].

These types of cells reproduce from differentiated lineages suitable for the tissue in which they are present and are therefore called multipotent or unipotent (figure 3.5).

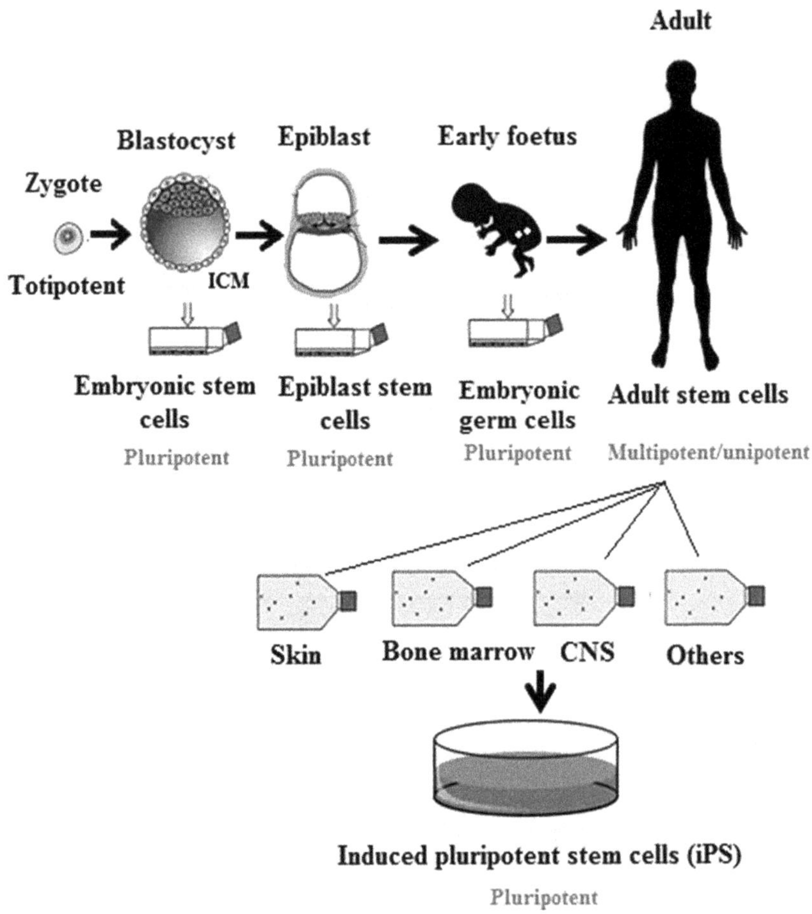

Figure 3.5. Induced pluripotent stem cells (iPSC).

Current research allows reprogramming of different cells to pluripotent and other cells. This can be achieved through their treatment with certain factors. The iPSC breakthrough has unlocked extraordinary prospects in the healthcare industry. Current research is mainly focused on iPSC applications in human disease modeling and stem cell treatment research [106–118]. As mentioned above, iPSCs are reprogrammed cells which are derived from somatic cells by treating them with defined transcription factors [106–118]. Such types of cells have the distinctive characteristics of self-renewal and differentiation to several types of cell lineage. Thus such types of cells can be utilized in the replacement of embryonic stem cells to surmount the different ethical concerns associated with the use of embryos [106–118]. Promising results have encouraged researchers worldwide to set up more effective procedures for iPSC generation. iPSC development requires great understanding, in particular related to the primary mechanisms involved in their development and use in treatment [106–118]. Huge amounts of data in the form of extensive reports are available which show the potential roles of various factors in developing iPSC generation methods. The reprogramming of somatic cells involves several molecules, such as DNA modifying agents (DNA methyl transferases), NANOG, miRNAs, etc [106–118]. These molecules and the associated reprogramming mechanisms require more attention to determine an unambiguous relationship between cell type and therapeutic response. In addition to their many treatment applications, iPSCs can also be utilized in studying the molecular mechanisms of several diseases [106–118]. Several disorders have now been modeled using iPSCs to develop a good understanding of the causes of the disease. This understanding can be further utilized to develop effective treatments against various diseases. Moreover, iPSCs are not utilized for the generation of patient-specific cells. These cells can be further utilized to transplant cells at a site of injury or degeneration. The utilization of iPSCs has now been expanded into engraftment processes, where iPSCs are utilized to reduce the risks of immune rejection as patient-specific cells can be used for transplantation. Moreover, this technology also provides several benefits over other similar techniques, such as animal models. iPSCs are also utilized to test toxic compounds and new drugs and their effects on the human system [106–118].

3.7.1 Stem cell applications in regenerative medicine and disease therapeutics

Regenerative medicine is an emerging branch of medical science that involves the regeneration of specific tissues and/or organs for an individual with severe injuries or chronic disease, where the potential outcomes of the individual's own regenerative responses are poor [83]. Currently, the tissues and organs which are donated are not meeting the demand for transplants, thus there is a need for an alternative procedure to meet the requirements of diseased and aged populations. Stem cells have enormous infinite cell division potential, which can be utilized in regenerative medicine to repair tissue and organ abnormalities due to [84]:

- congenital defects,
- diseases,
- age related effects.

Stem cells offer a treatment platform for all the tissue and organ systems of the body. These cells also play an important role in development and tissue repair processes. Based on the type of lineage reprogramming, also called metaplasia or transdifferentiation, where stem cells are converted from one cell type to another cell type, stem cells are divided into four types [85]:

- unipotent,
- multipotent,
- pluripotent,
- totipotent.

Totipotent cells can give rise to a whole organism through the process of transdifferentiation, and the best example of totipotent cells is the zygote, as the embryonic inner cell mass (blastocysts, before implantation within the uterus) is only a pluripotent cell lineage as the cells differentiate the three germ layers but do not differentiate into the cells of extra-embryonic tissue [86]. Pluripotency-associated genes, also called reprogramming factors or transcription factors, can be utilized to reprogram or produce iPSCs, which are functionally similar to embryonic stem cells. These factors include Oct3/4, Sox2, c-Myc, Klf4 and NANOG, also called the 'Yamanaka factors', and they have transformed regenerative medicine because they are responsible for the stemness (the essential characteristic of a stem cell that differentiates it from ordinary cells) and trans-differentiation potential of embryonic, extra-embryonic, fetal and adult stem cells [85–87]. However, the transcription factors (factors that control the expression of several genes) used to induce stemness are also overexpressed in cancer, which can result in cancer progression and pathogenesis. Not only overexpression but also mutations and loss of expression can result in cancer. When administered into the tissue, ES cells and iPSCs cause teratoma formation (i.e. tumor-like formations containing tissues belonging to all three germ layers) which raises the question of the safety of treatments based on iPSC derivatives. As mentioned above, lineage reprogramming involves specialized changes of one cell type into another without entering a pluripotent state. This process involves an ectopic expression (abnormal gene expression in a cell where the gene is not usually expressed) of transcription factors. In the case of reprogramming, there is no risk of subsequent teratoma-genesis. Pluripotent stem cells (iPSCs) [85–87] result from the ectopic expression or functional restoration of endogenous pluripotency factors including OCT4, SOX2, NANOG and MYC. These transcription factors are responsible for the epigenetic transformation of terminally differentiated cells into ES cell-like cells [85–87]. All stem cells can, depending on their differentiation potential and regenerative applications, be classified into six categories:

- ES cells,
- tissue specific progenitor stem cells (TSPSCs),
- mesenchymal stem cells (MSCs),
- umbilical cord stem cells (UCSCs),
- bone marrow stem cells (BMSCs),
- iPSCs.

The transplantation of stem cells can be:

- *Autologous.* The patient's own blood-forming stem cells are collected.
- *Allogenic.* A donor stem cell transplant replaces bone marrow that is no longer working properly with healthy stem cells from a donor.
- *Xenogeneic.* Stem cells from different species are transplanted.
- *Autograft.* Transplantation of cells, tissues or organs between sites within the same individual.
- *Syngeneic.* For the induction of tissue regeneration.

To prevent host–graft rejections, a tissue typing approach is utilized in which the tissues of a potential donor and recipient are examined properly for compatibility before transplantation, using the following methods. Human leukocyte antigens

(HLAs) are examined for tissue and organ transplants as well as the use of immune suppressants [88]. HLAs are a group of genes that encode the proteins responsible for identification of agents foreign to the immune system. These proteins are called 'self-markers', as they are present over the surface of all cells and instruct the immune system not to generate a response unless activated.

The approaches in regenerative medicine that can be utilized for physiological restoration of damaged tissue and organs involve tissue engineering and its applications in cellular transplantation, material science and microengineering for the development of organoids. This technology involves a 3D biodegradable scaffold, a 3D substrate that serves as a template for the tissue [89, 90]. Ideally scaffolds should have the following characteristics [91]:

- Nonimmunogenic for the host.
- Mimic the mechanics of the target tissue.
- Not require systemic immune suppressant.
- Support angiogenesis and neovascularisation for suitable tissue perfusion.
- Support cell adhesion and ingrowths.

The *ex vivo* expansion of stem cells for clinical use is a valuable and safe approach to increase stem cell numbers so as to increase engraftment and reduce the chances of morbidity from infection. The *ex vivo* expansion of stem cells results in an increase in the stem cell count [92, 93]. Stem cell-based regenerative therapy is considered fruitful when the transplanted stem cells [94]:

- Survive for longer periods.
- Proliferate efficiently.
- Differentiate efficiently in a site-specific manner.
- Integrate efficiently into the host circulatory system.

3.7.2 Other applications

Major developments in the field of stem cell therapy are discussed in the following.

3.7.2.1 Reprogramming technology

The breakthrough of iPSCs opened the way for reprogramming technology (figure 3.6). By means of this technology we can alter the cell destiny by over-expression of master transcriptional factors (Oct3/4, Sox2, c-Myc and Klf4). Various applications of this technology have recently been explored, such as the development of neuronal cells directly induced from somatic cells (figure 3.6). It has recently been shown that the overexpression of transcriptional factors in neuronal cells can help change the cell destiny of endogenous astroglia to neuronal cells under *in vivo* conditions [95–105].

Thus by means of this technology it is possible to convert differentiated cells of one type into cells of a completely different origin and function. This can be achieved either directly or via the production of induced pluripotent stem cells.

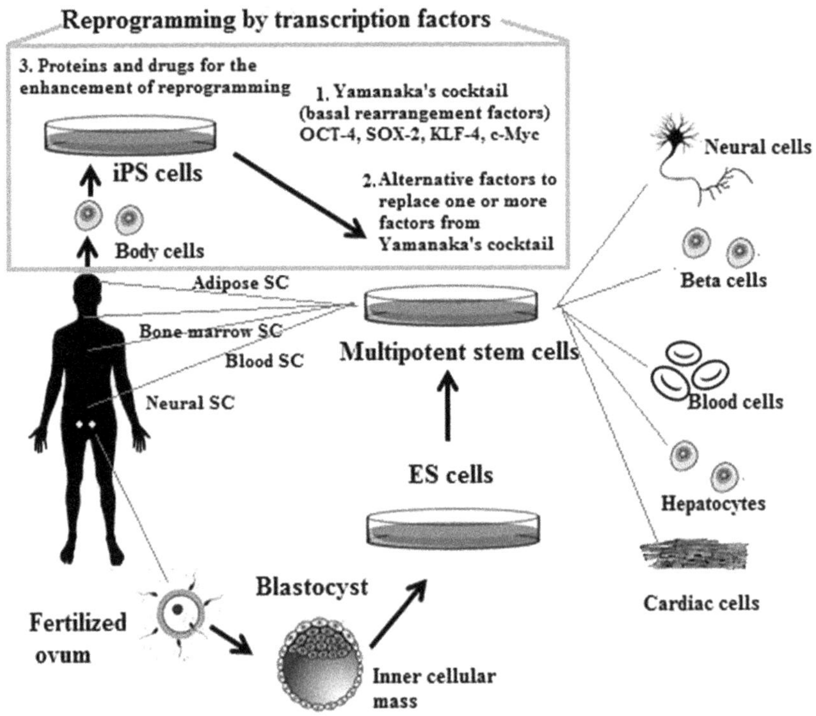

Figure 3.6. Cellular reprogramming by transcription factors.

3.7.2.2 Targeted genome editing

CRISPR gene editing technology is a recent innovation in the field. Prokaryotes have developed numerous defense mechanisms to protect themselves, in particular from viruses. The prokaryotic immune system contains clustered, regularly spaced, short, palindromic repeats (CRISPR) and their associated proteins (Cas) [95–105]. This technology allows researchers to edit parts of the genome by removing, adding or altering sections of the DNA sequence. It is now the simplest, most useful and accurate method of genetic manipulation. The system of CRISPR–Cas9 consists of two important molecules that introduce a mutation into the DNA. These are:

- The enzyme known as Cas9 nuclease. The role of this enzyme is to cut or slice the two strands of DNA at a particular site in the genome. This enzyme called 'molecular scissors' as it cuts in such a manner that pieces of DNA can then be added or removed [95–105].
- The RNA fragment known as guide RNA (gRNA). This guide consists of a small fragment of pre-designed RNA sequence (about 20 bases long) present within a longer RNA scaffold or support. The support part binds to the DNA and the pre-designed sequence directs or guides the Cas9 nuclease enzyme to the right part of the genome to make a cut at a suitable point in the genome [95–105].

CRISPR–Cas9 work together collaboratively in the immune system of prokaryotic cells, conferring resistance to foreign genetic elements [95–105]. In the last few years the CRISPR–Cas9 system has been used to edit genomes. CRISPR–Cas9 works by transporting Cas9 nuclease complexed with a synthetic guide RNA. This delivery of the enzyme with an RNA guide allows the cell to make a cut at the desirable site in the genome and new genes can be added. CRISPR–Cas9 is considered to be the most successful scientific tool that can edit the genome, as it is the most efficient, inexpensive and user-friendly genome editing technology. Thus the entire process is based on the concept of genome editing, which can be accomplished by engineered nucleases that can cut an organism's DNA at a desirable site [95–105]. These sites can be genetically modified and later repaired by homologous recombination or non-homologous end joining. The early established gene editing tools are Zinc Fingers, TAL effectors and TALEN. CRISPR–Cas9 has proved to be more successful than these previous inventions and was acknowledged as *Science*'s 2015 'breakthrough of the year' discovery [95–105].

3.7.2.3 Large-scale production

Pluripotent stem cells can be cultured *in vitro* and differentiated into all cell types of the human body. Thus they represent highly promising cell sources for biomedical applications such as cell therapies, tissue engineering, and drug discovery. Several efficient, defined, scalable, and good manufacturing practice-compatible 3D systems for the production of human pluripotent stem cells and their progeny are available. The ease of use and flexible scalability of these systems make them suitable for numerous applications from the laboratory toward the clinic.ESCs and iPSCs can now be cultured on a large scale using suspension culture, and large numbers of human PSCs can be produced in fully controlled bioreactors in defined culture media. In addition, sequential inhibition and activation of molecular differentiation pathways has allowed a targeted, more robust and efficient differentiation of human ESCs and iPSCs.

3.7.2.4 Improved yield

Yielding highly enriched preparations of well-defined cell lineages at clinically required dimensions is a challenge, however, techniques such as co-cultures are about to achieve production in a suitably short period of time [119, 120].

For most of the applications of stem cells, targeted specification is required, in particular for distinct cellular subpopulations. Additionally, proper cell maturation is also required. However, certain challenges such as low survival rates and inadequate functional integration of cellular transplants still need to be overcome [119, 120].

References

[1] Keung A J, Healy K E, Kumar S and Schaffer D V 2010 Biophysics and dynamics of natural and engineered stem cell microenvironments *Interdiscip. Rev. Syst. Biol. Med.* **2** 49–64

[2] Willerth S M and Sakiyama-Elbert S E 2008 Combining stem cells and biomaterial scaffolds for constructing tissues and cell delivery *StemBook* (Cambridge, MA: Harvard Stem Cell Institute)
[3] Lee J, Cuddihy M J and Kotov N A 2008 Three-dimensional cell culture matrices: state of the art *Tissue Eng.* B **14** 61–86
[4] Ahmed T A, Dare E V and Hincke M 2008 Fibrin: a versatile scaffold for tissue engineering applications *Tissue Eng.* B **14** 199–215
[5] Breen A, O'Brien T and Pandit A 2009 Fibrin as a delivery system for therapeutic drugs and biomolecules *Tissue Eng.* B **15** 201–14
[6] Willerth S M, Faxel T E, Gottlieb D I and Sakiyama-Elbert S E 2007 The effects of soluble growth factors on embryonic stem cell differentiation inside of fibrin scaffolds *Stem Cells* **25** 2235–44
[7] Shufaro Y and Reubinoff B E 2004 Therapeutic applications of embryonic stem cells *Best Pract. Res. Clin. Obstet. Gynaecol.* **18** 909–27
[8] Vazin T and Freed W J 2010 Human embryonic stem cells: derivation, culture, and differentiation, a review *Restor. Neurol. Neurosci.* **28** 589–603
[9] Chen S *et al* 2009 A small molecule that directs differentiation of human ESCs into the pancreatic lineage *Nat. Chem. Biol.* **5** 258–65
[10] Harding S E, Ali N N, Brito-Martins M and Gorelik J 2007 The human embryonic stem cell-derived cardiomyocyte as a pharmacological model *Pharmacol. Ther.* **113** 341–53
[11] Agarwal S, Holton K L and Lanza R 2008 Efficient differentiation of functional hepatocytes from human embryonic stem cells *Stem Cells* **26** 1117–27
[12] Klimanskaya I, Rosenthal N and Lanza R 2008 Derive and conquer: sourcing and differentiating stem cells for therapeutic applications *Nat. Rev. Drug Discov.* **7** 131–42
[13] Sachlos E *et al* 2012 Identification of drugs including a dopamine receptor antagonist that selectively target cancer stem cells *Cell* **149** 1284–97
[14] Takahashi K and Yamanaka S 2006 Induction of pluripotent stem cells from mouse embryonic and adult fibroblast cultures by defined factors *Cell* **126** 663–76
[15] Grskovic M, Javaherian A, Strulovici B and Daley G Q 2011 Induced pluripotent stem cells opportunities for disease modeling and drug discovery *Nat. Rev. Drug Discov.* **10** 915–29
[16] Bilic J and Izpisua Belmonte J C 2012 Induced pluripotent stem cells versus embryonic stem cells: close enough or yet too far apart? *Stem Cells* **30** 33–41
[17] Zhu H *et al* 2011 Investigating monogenic and complex diseases with pluripotent stem cells *Nat. Rev. Genet.* **12** 266–75
[18] Kim K *et al* 2010 Epigenetic memory in induced pluripotent stem cells *Nature* **467** 285–90
[19] Zhu D Y and Lou Y J 2005 Inducible effects of icariin, icaritin, and desmethylicaritin on directional differentiation of embryonic stem cells into cardiomyocytes *in vitro Acta Pharm. Sin.* **26** 477–85
[20] Zhu D Y and Lou Y J 2006 Icariin-mediated expression of cardiac genes and modulation of nitric oxide signaling pathway during differentiation of mouse embryonic stem cells into cardiomyocytes *in vitro Acta Pharm. Sin.* **27** 311–20
[21] Zhu D Y *et al* 2008 MAPEG expression in murine embryonic stem cell-derived hepatic tissue system *Stem Cell Dev.* **17** 775–84
[22] Hao J J *et al* 2008 Dorsomorphin, a selective small molecule inhibitor of BMP signaling, promotes cardiomyogenesis in embryonic stem cells *PLoS One* **3** 1–8

[23] Ogawa K *et al* 2007 Activin-nodal signaling is involved in propagation of mouse embryonic stem cells *J. Cell Sci.* **120** 55–65
[24] Peters A K, Steemans M, Hansen E, Mesens N, Verheyen G R and Vanparys P 2008 Evaluation of the embryotoxic potency of compounds in a newly revised high throughput embryonic stem cell test *Toxicol. Sci.* **105** 342–50
[25] Elkabetz Y, Panagiotakos G, Shamy G A, Socci N D, Tabar V and Studer L 2008 Human ES cell-derived neural rosettes reveal a functionally distinct early neural stem cell stage *Gene Dev.* **22** 152–65
[26] Hakuno D, Takahashi T, Lammerding J and Lee R T 2005 Focal adhesion kinase signaling regulates cardiogenesis of embryonic stem cells *J. Biol. Chem.* **280** 39534–44
[27] Pan G and Thomson J A 2007 NANOG and transcriptional networks in embryonic stem cell pluripotency *Cell Res.* **17** 42–9
[28] Sampath P *et al* 2008 A hierarchical network controls protein translation during murine embryonic stem cell self-renewal and differentiation *Cell Stem Cell* **2** 448–60
[29] Babale Y *et al* 2007 Analysis of Oct4-dependent transcriptional networks regulating self-renewal and pluripotency in human embryonic stem cells *Stem Cells* **25** 500–10
[30] Peerani R *et al* 2007 Niche-mediated control of human embryonic stem cell self-renewal and differentiation *EMBO J.* **26** 4744–55
[31] Morag H, Stewart M H, Bendall S C and Bhatia M 2008 Deconstructing human embryonic stem cell cultures: niche regulation of self-renewal and pluripotency *J Mol. Med.* **86** 875–86
[32] Yao S *et al* 2006 Long-term self-renewal and directed differentiation of human embryonic stem cells in chemically defined conditions *Proc. Natl Acad. Sci. USA* **103** 6907–12
[33] Bouhona I A, Kato H, Chandran S and Allen N D 2005 Neural differentiation of mouse embryonic stem cells in chemically defined medium *Brain. Res. Bull.* **68** 62–75
[34] Firestone A J and Chen J K 2010 Controlling destiny through chemistry: small-molecule regulators of cell fate *ACS Chem. Biol.* **5** 15–34
[35] Krencik R and Zhang S C 2006 Stem cell neural differentiation: a model for chemical biology *Curr. Opin. Chem. Biol.* **10** 592–7
[36] Xu Y, Shi Y and Ding S 2008 A chemical approach to stem-cell biology and regenerative medicine *Nature* **453** 338–44
[37] Brito-Martins M, Harding S E and Ali N N 2008 β1- and 2-adrenoceptor responses in cardiomyocytes derived from human embryonic stem cells: comparison with failing and non-failing adult human heart *Br. J. Pharmacol.* **153** 751–9
[38] Moore K A and Lemischka I R 2006 Stem cells and their niches *Science* **311** 1880–5
[39] Huang X, Cho S and Spangrude G J 2007 Hematopoietic stem cells: generation and self-renewal *Cell Death Differ.* **14** 1851–9
[40] Clevers H 2016 Modeling development and disease with organoids *Cell* **165** 1586–97
[41] Rheinwald J G and Green H 1975 Serial cultivation of strains of human epidermal keratinocytes: the formation of keratinizing colonies from single cells *Cell* **6** 331–43
[42] Barrandon Y and Green H 1987 Three clonal types of keratinocyte with different capacities for multiplication *Proc. Natl Acad. Sci. US A* **84** 2302–6
[43] Senoo M, Pinto F, Crum C P and McKeon F 2007 p63 Is essential for the proliferative potential of stem cells in stratified epithelia *Cell* **129** 523–36
[44] Yang A *et al* 1999 p63 is essential for regenerative proliferation in limb, craniofacial and epithelial development *Nature* **398** 714–8

[45] Pellegrini G *et al* 2001 p63 identifies keratinocyte stem cells *Proc. Natl Acad. Sci. USA* **98** 3156–61
[46] Viganò M A and Mantovani R 2007 Hitting the numbers: the emerging network of p63 targets *Cell Cycle* **6** 233–9
[47] Romano R A and Sinha S 2011 Dynamic life of a skin keratinocyte: an intimate tryst with the master regulator p63 *Indian. J. Exp. Biol.* **49** 721–31
[48] O'Connor N E, Mulliken J B, Banks-Schlegel S, Kehinde O and Green H 1981 Grafting of burns with cultured epithelium prepared from autologous epidermal cells *Lancet* **317** 75–8
[49] Gallico G G 3rd, O'Connor N E, Compton C C, Kehinde O and Green H 1984 Permanent coverage of large burn wounds with autologous cultured human epithelium *N. Engl. J. Med.* **311** 448–51
[50] Kim J W, Lee J H, Lyoo Y S, Jung D I and Park H M 2013 The effects of topical mesenchymal stem cell transplantation in canine experimental cutaneous wounds *Vet. Dermatol.* **24** 242–53
[51] Blais M, Parenteau-Bareil R, Cadau S and Berthod F 2013 Tissue-engineered skin and nerve regeneration in burn treatment *Stem Cells Transl. Med.* **2** 545–51
[52] Pouliot R, Larouche D, Auger F A, Juhasz J, Xu W, Li H and Germain L 2002 Reconstructed human skin produced *in vitro* and grafted on athymic mice *Transplantation* **73** 1751–7
[53] Martínez-Santamaría L, Guerrero-Aspizua S and Del Río M 2012 Skin bioengineering: preclinical and clinical applications *Actas Dermosifiliogr.* **103** 5–11
[54] Reid M J, Currie L J, James S E and Sharpe J R 2007 Effect of dermal substitute, cultured keratinocytes and split thickness skin graft on wound contraction *Wound Repair Regen.* **15** 889–96
[55] Pellegrini G, Traverso C E, Franzi A T, Zingirian M, Cancedda R and De Luca M 1997 Long-term restoration of damaged corneal surfaces with autologous cultivated corneal epithelium *Lancet* **349** 990–3
[56] Rama P, Matuska S, Paganoni G, Spinelli A, De Luca M and Pellegrini G 2010 Limbal stem-cell therapy and long-term corneal regeneration *N. Engl. J. Med.* **363** 147–55
[57] Bisson F *et al* 2013 Irradiated human dermal fibroblasts are as efficient as mouse fibroblasts as a feeder layer to improve human epidermal cell culture lifespan *Int. J. Mol. Sci.* **14** 4684–704
[58] Barrandon Y and Green H 1987 Three clonal types of keratinocyte with different capacities for multiplication *Proc. Natl Acad. Sci. USA* **84** 2302–6
[59] Bonfanti P, Claudinot S, Amici A W, Farley A, Blackburn C C and Barrandon Y 2010 Microenvironmental reprogramming of thymic epithelial cells to skin multipotent stem cells *Nature* **466** 978–82
[60] Sun T T, Bonitz P and Burns W H 1984 Cell culture of mammalian thymic epithelial cells: growth, structural, and antigenic properties *Cell Immunol.* **83** 1–13
[61] Larsson H M, Gorostidi F, Hubbell J A, Barrandon Y and Frey P 2014 Clonal, self-renewing and differentiating human and porcine urothelial cells, a novel stem cell population *PLoS One* **9** e90006
[62] Kumar P A *et al* 2011 Distal airway stem cells yield alveoli *in vitro* and during lung regeneration following H1N1 influenza infection *Cell* **147** 525–38
[63] Zuo W *et al* 2015 p63(+)Krt5(+) distal airway stem cells are essential for lung regeneration *Nature* **517** 616–20

[64] Yamamoto Y *et al* 2016 Mutational spectrum of Barrett's stem cells suggests paths to initiation of a precancerous lesion *Nat. Commun.* **7** 10380
[65] Liu X *et al* 2012 ROCK inhibitor and feeder cells induce the conditional reprogramming of epithelial cells *Am. J. Pathol.* **180** 599–607
[66] Chapman S, Liu X, Meyers C, Schlegel R and McBride A A 2010 Human keratinocytes are efficiently immortalized by a Rho kinase inhibitor *J. Clin. Invest.* **120** 2619–26
[67] Yuan H *et al* 2012 Use of reprogrammed cells to identify therapy for respiratory papillomatosis *N. Engl. J. Med.* **367** 1220–7
[68] Karthaus W R *et al* 2014 Identification of multipotent luminal progenitor cells in human prostate organoidcultures *Cell* **159** 163–75
[69] Eiraku M *et al* 2011 Self-organizing optic-cup morphogenesis in three-dimensional culture *Nature* **472** 51–6
[70] Takebe T *et al* 2013 Vascularized and functional human liver from an iPSC-derived organ bud transplant *Nature* **499** 481–4
[71] Barker N *et al* 2007 Identification of stem cells in small intestine and colon by marker gene Lgr5 *Nature* **449** 1003–7
[72] Sato T *et al* 2009 Single Lgr5 stem cells build crypt-villus structures *in vitro* without a mesenchymal niche *Nature* **459** 262–5
[73] Sato T *et al* 2011 Long-term expansion of epithelial organoids from human colon, adenoma, adenocarcinoma, and Barrett's epithelium *Gastroenterology* **141** 1762–72
[74] Jung P *et al* 2011 Isolation and *in vitro* expansion of human colonic stem cells *Nat. Med.* **17** 1225–7
[75] Huch M *et al* 2013 Unlimited *in vitro* expansion of adult bi-potent pancreas progenitors through the Lgr5/R-spondin axis *EMBO J.* **32** 2708–21
[76] Huch M *et al* 2015 Long-term culture of genome-stable bipotent stem cells from adult human liver *Cell* **160** 299–312
[77] McGranahan N and Swanton C 2015 Biological and therapeutic impact of intratumor heterogeneity in cancer evolution *Cancer Cell* **27** 15–26
[78] DeRose Y S *et al* 2011 Tumor grafts derived from women with breast cancer authentically reflect tumor pathology, growth, metastasis and disease outcomes *Nat. Med.* **17** 1514–20
[79] Liu E T, Bult C J and Shultz L D 2016 Patient-derived tumor xenografts: why now? *JAMA Oncol.* **2** 567–8
[80] Goers L, Freemont P and Polizzi K M 2014 Co-culture systems and technologies: taking synthetic biology to the next level *J. R. Soc. Interface* **11** 20140065
[81] Arias A M 2008 *Drosophila melanogaster* and the development of biology in the 20th century *Methods Mol. Biol.* **420** 1–25
[82] Lajtha L G 1979 Stem cell concepts *Nouv. Rev. Fr. Hematol.* **21** 59–65
[83] Mason C and Dunnill P 2008 A brief definition of regenerative medicine *Regen. Med.* **3** 1–5
[84] Fortier L A 2005 Stem cells: classifications, controversies, and clinical applications *Vet. Surg.* **34** 415–23
[85] Meissner A, Wernig M and Jaenisch R 2007 Direct reprogramming of genetically unmodified fibroblasts into pluripotent stem cells *Nat. Biotechnol.* **25** 1177–81
[86] Yu J *et al* 2007 Induced pluripotent stem cell lines derived from human somatic cells *Science* **318** 1917–20
[87] Thomson M, Liu S J, Zou L N, Smith Z, Meissner A and Ramanathan S 2011 Pluripotency factors in embryonic stem cells regulate differentiation into germ layers *Cell* **145** 875–89

[88] Petersdorf E W, Malkki M, Gooley T A, Martin P J and Guo Z 2007 MHC haplotype matching for unrelated hematopoietic cell transplantation *PLoS Med.* **4** e8
[89] Leferink A M, Chng Y C, van Blitterswijk C A and Moroni L 2015 Distribution and viability of fetal and adult human bone marrow stromal cells in a biaxial rotating vessel bioreactor after seeding on polymeric 3D additive manufactured scaffolds *Front. Bioengin. Biotechnol.* **3** 169
[90] Gubareva E A *et al* 2016 Orthotopic transplantation of a tissue engineered diaphragm in rats *Biomaterials* **77** 320–35
[91] Garzón I *et al* 2012 Evaluation of the cell viability of human Wharton's Jelly stem cells for use in cell therapy *Tiss. Eng.* C **18** 408–19
[92] Thompson P A *et al* 2016 Double umbilical cord blood transplant is effective therapy for relapsed or refractory Hodgkin lymphoma *Leuk. Lymphoma* **57** 1607–15
[93] Nathamgari S S *et al* 2015 Isolating single cells in a neurosphere assay using inertial microfluidics *Lab Chip* **15** 4591–7
[94] Ning B, Cheuk D K, Chiang A K, Lee P P, Ha S Y and Chan G C 2015 Autologous cord blood transplantation for metastatic neuroblastoma *Pediatr. Transplant.* **20** 290–6
[95] Yamashita T and ABE K 2016 Recent progress in cell reprogramming technology for cell transplantation therapy *Neurol. Med.-Chir.* **56** 97–101
[96] Barrangou R, Fremaux C, Deveau H, Richards M, Boyaval P, Moineau S, Romero D A and Horvath P 2007 CRISPR provides acquired resistance against viruses in prokaryotes *Science* **315** 1709–12
[97] Haft D H, Selengut J, Mongodin E F and Nelson K E 2005 A guild of 45 CRISPR-associated (Cas) protein families and multiple CRISPR/Cas subtypes exist in prokaryotic genomes *PLoS Comput. Biol.* **1** e60
[98] Makarova K S, Grishin N V, Shabalina S A, Wolf Y I and Koonin E V 2006 A putative RNA-interference-based immune system in prokaryotes: computational analysis of the predicted enzymatic machinery, functional analogies with eukaryotic RNAi, and hypothetical mechanisms of action *Biol. Direct.* **1** 7
[99] Mojica F J, Diez-Villasenor C, Garcia-Martinez J and Soria E 2005 Intervening sequences of regularly spaced prokaryotic repeats derive from foreign genetic elements *J. Mol. Evol.* **60** 174–82
[100] Makarova K S *et al* 2015 An updated evolutionary classification of CRISPR–Cas systems *Nat. Rev. Microbiol.* **13** 722–36
[101] Makarova K S *et al* 2011 Evolution and classification of the CRISPR–Cas systems *Nat. Rev. Microbiol.* **9** 467–77
[102] Shmakov S *et al* 2015 Discovery and functional characterization of diverse class 2 CRISPR–Cas systems *Mol. Cell.* **60** 385–97
[103] Fineran P C and Charpentier E 2012 Memory of viral infections by CRISPR–Cas adaptive immune systems: acquisition of new information *Virology* **434** 202–9
[104] Rath D, Amlinger L, Rath A and Lundgren M 2015 The CRISPR–Cas immune system: biology, mechanisms and applications *Biochimie* **117** 119–28
[105] Nunez J K, Kranzusch P J, Noeske J, Wright A V, Davies C W and Doudna J A 2014 Cas1–Cas2 complex formation mediates spacer acquisition during CRISPR–Cas adaptive immunity *Nat. Struct. Mol. Biol.* **21** 528–34
[106] Takahashi K and Yamanaka S 2013 Induced pluripotent stem cells in medicine and biology *Development.* **140** 2457–61

[107] Singh V K, Kalsan M, Kumar N, Saini A and Chandra R 2015 Induced pluripotent stem cells: applications in regenerative medicine, disease modeling, and drug discovery *Front. Cell Dev. Biol.* **3** 2
[108] Nadig R R 2009 Stem cell therapy—hype or hope? A review *J. Conserv. Dent.* **12** 131–8
[109] Zaehres H, Kim J B and Schöler H R 2010 Induced pluripotent stem cells *Methods Enzymol.* **476** 309–25
[110] Park I H, Zhao R, West J A, Yabuuchi A, Huo H, Ince T A, Lerou P H, Lensch M W and Daley G Q 2008 Reprogramming of human somatic cells to pluripotency with defined factors *Nature* **451** 141–6
[111] Ebben J D, Zorniak M, Clark P A and Kuo J S 2011 Introduction to induced pluripotent stem cells: advancing the potential for personalized medicine *World Neurosurg* **76** 270–5
[112] Yamanaka S 2012 Induced pluripotent stem cells: past, present, and future *Cell Stem Cell* **10** 678–84
[113] Song L and Goldman E 2015 Induced pluripotent stem cells are induced pluripotent stem cell-like cells *J. Biomed. Res.* **29** 1–2
[114] Stadtfeld M and Hochedlinger K 2010 Induced pluripotency: history, mechanisms, and applications *Genes Dev.* **24** 2239–63
[115] Okita K and Yamanaka S 2011 Induced pluripotent stem cells: opportunities and challenges *Philos. Trans. R. Soc. Lond.* B **366** 2198–207
[116] Saha K and Jaenisch R 2009 Technical challenges in using human induced pluripotent stem cells to model disease *Cell Stem Cell* **5** 584–95
[117] Zhao J, Jiang W J, Sun C, Hou C Z, Yang X M and Gao J G 2013 Induced pluripotent stem cells: origins, applications, and future perspectives *J. Zhejiang Univ. Sci.* B **14** 1059–69
[118] Kumar D, Talluri T R, Anand T and Kues W A 2015 Induced pluripotent stem cells: mechanisms, achievements and perspectives in farm animals *World J. Stem Cells* **7** 315–28
[119] Bhatia S 2018 *Introduction to Pharmaceutical Biotechnology: Enzymes, Proteins and Bioinformatics* vol 2 (Bristol: IOP Publishing)
[120] Bhatia S 2018 *Introduction to Pharmaceutical Biotechnology: Basic Techniques and Concepts* vol 1 (Bristol: IOP Publishing)

IOP Publishing

Introduction to Pharmaceutical Biotechnology, Volume 3
Animal tissue culture and biopharmaceuticals
Saurabh Bhatia, Tanveer Naved and Satish Sardana

Chapter 4

Culture media for animal cells

4.1 Introduction

Cell culture technology is considered one of the major breakthroughs in the life sciences. This technology involves the removal of cells, tissues or organs from an animal or plant and their subsequent placement into an artificial environment conducive to their survival and/or proliferation [1–7]. The optimal environmental conditions required for cell growth are a controlled temperature, a substrate for cell attachment, an appropriate growth medium, and an incubator that maintains correct pH and osmolality [1–7]. The selection of the growth medium for *in vitro* cultivation is considered as the most important and crucial step in cell culture. The medium can be the substrate, a platform which is required for the growth of some cells under *in vitro* conditions. The growth or culture medium can thus be liquid or solid [1–7]. Cell culture media generally comprise an appropriate source of energy and compounds which regulate the cell cycle. A typical culture medium is composed of a complement of amino acids, vitamins, inorganic salts, glucose and serum as a source of growth factors, hormones and attachment factors. In addition to nutrients, the medium also helps maintain pH and osmolality. Since blood meets the nutritional requirements of mammalian cells, animal cell culture medium should be prepared in such a way that it contains all the components present in blood [1–7]. Generally, the type of cell considered for culture and the purpose of the culture determine the type of medium. Broadly, culture media are divided into two classes: natural and artificial media.

4.1.1 Natural media

Previously, natural media were derived from different resources, such as plasma, serum, lymph, amniotic fluid, ascetic and pleural fluids, and aqueous humor from the eyes. Sterilization procedures for these biological fluids have been developed, as have methods to determine their toxicity before utilization. Simulated body fluid, a solution with an ion concentration almost similar to human blood plasma, has been extensively used for the bioactivity assessment of biomaterials. Various researchers have suggested approaches to overcome the shortcomings of conventional simulated body fluid. Natural media also include tissue extracts in the form of chick embryo extract [8, 9] and liver, bone marrow and spleen extracts. Different types of natural media are listed in figure 4.1.

4.1.2 Artificial media

A medium with a designed composition is called an artificial medium [10]. Different types of artificial media are described in figure 4.2. The minimum requirements for artificial/synthetic media are:

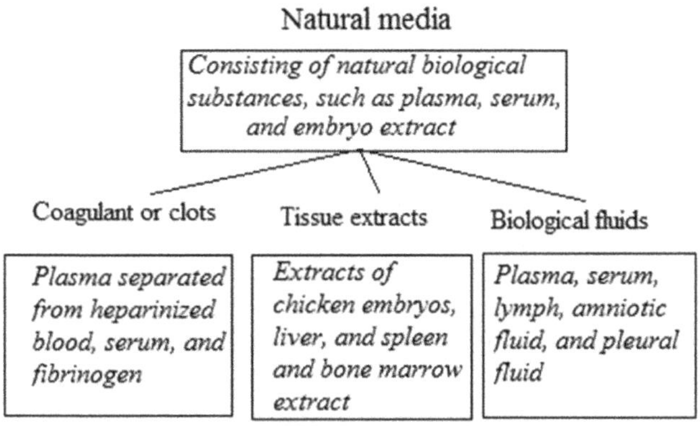

Figure 4.1. Various types of natural media.

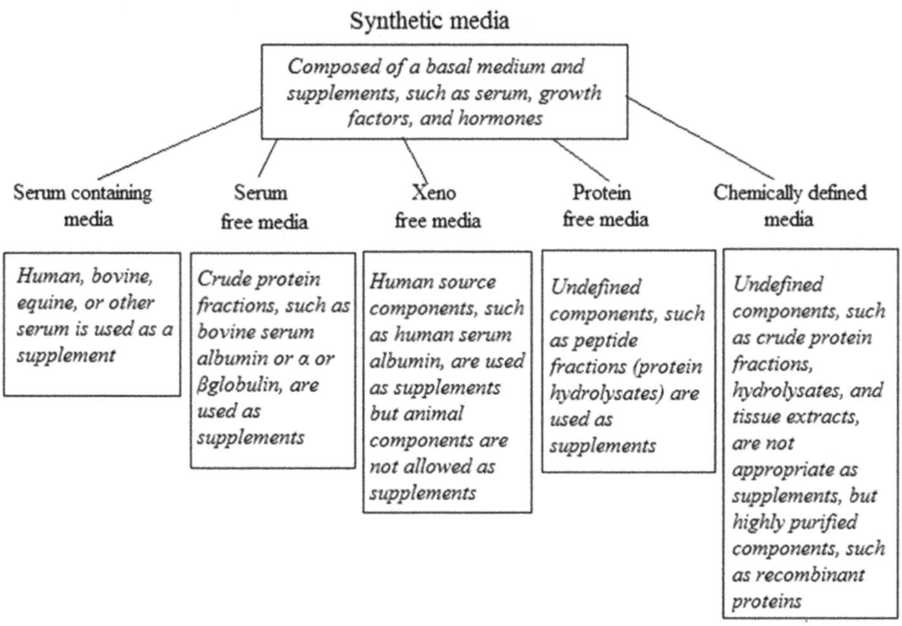

Figure 4.2. Different types of artificial media.

- The medium must offer all the nutrients to the cell in a sustainable manner.
- The medium must be isotonic and sterilized before utilization.
- The medium has to offer optimum pH (around 7).

In controlled conditions, a balanced salt solution is generally used to maintain biological pH and osmolality in a culture under *in vitro* conditions. To encourage growth, various essential and non-essential elements in the form of organic salts, growth factors (vitamins, amino acids, etc) and natural supplements are added. Supplementation of serum is required in most cases, however, currently several researchers are using serum-free media.

4.2 Physical and chemical properties of culture media

The physical and chemical properties of culture media and their effects are described in table 4.1.

4.2.1 pH

The desirable pH range for a culture medium is pH 6.4–8.3. The metabolism and growth of cultured cells are significantly influenced by pH variation in the medium, in particular bicarbonate-buffered media. Such pH variations can be overcome by using suitable combinations of certain organic buffers, as mentioned in table 4.1, so that the pH of a culture can be stabilized by appropriate combinations of two or three buffers. The initial alkalization can be controlled by using buffers, however, the metabolic acidification (due to the constant production of metabolites) is usually less than 0.4 pH units, excluding heavy cultures [11].

A biocarbonate buffer system controls the pH of the culture medium (figure 4.3). Sodium bicarbonate, when dissolved in water, results in the formation of sodium and bicarbonate ions. The bicarbonate ions reacts with hydrogen ions present in the solution to form carbonic acid, which ultimately forms carbon dioxide and water. So, based on the Henderson–Hasselbalch equation both reactions are in equilibrium with each other, i.e. the carbon dioxide present in the solution is in equilibrium with the carbon dioxide present in the gas phase, thus increasing the concentration of carbon dioxide in the gas phase enhances the concentration of carbon dioxide in the medium, and subsequently increases the carbonic acid concentration and decreases the pH. Conversely, if the carbon dioxide concentration in the gas phase is decreased, then the pH increases [11–15].

4.2.2 Oxygen

Oxygen is vital for aerobic respiration. Mammalian cells are well adapted to O_2 levels much lower than atmospheric conditions. The levels of O_2 in the environment outside the cell must be properly maintained to meet the requirements of cells. This is applicable for both *in vivo* and *in vitro* environments. To investigate the molecular and cellular responses, cell culture experiments have been performed under various O_2 levels, which can be achieved by range of technologies (e.g. gas-permeable technology) and instruments (e.g. gas-tight boxes and gas-controlled incubators) (see table 4.2). Tissue hypoxia is a common pathophysiological process. From the 19th

Table 4.1. Physicochemical parameters and their affects on culture media.

Parameters	General information	Solution
pH	The ideal pH range is 7–7.4. The indicator phenol red [11–15] is used for visible detection of pH variation in the medium in form of a change in color.	At pH 7.4: red At pH 7.0: orange At pH 6.5: yellow At pH 7.8: purple
CO_2, biocarbonate buffering	CO_2 in the medium is in a dissolved state, its concentration depends on temperature and atmospheric CO_2. CO_2 is supplied by carbonic acid and bicarbonate: $CO_2 + H_2O \leftrightarrow H_2CO_3 \leftrightarrow H^+ + HCO^-_3$. The above reaction confirms that HCO^-_3, CO_2 and pH are correlated. Thus by increasing CO_2 pH can be reduced and by addition of $NaHCO_3$ biocarbonate ions can be neutralized: $NaHCO_3 \leftrightarrow Na^+ + HCO^-_3$.	Bicarbonate is less toxic and inexpensive, and is often utilized. HEPES (hydroxyl ethyl piperazine 2-sulfonic acid) is more efficient then bicarbonate [11–15]. Pyruvate (by adding a high concentration of amino acids) is responsible for the endogenous production of CO_2 by the cells.
Oxygen	Most cells under *in vivo* conditions are dependent on oxygen for aerobic respiration. This can be achieved by supplying hemoglobin constantly to the tissue. However, culture cells under *in vitro* conditions are dependent on the dissolved oxygen concentration, which can be toxic to the culture if it increases as it may lead to the production of free radicals.	Free radical scavengers or antioxidants have to be added such as vitamins [C, E], glutathione, selenite, β-mercaptoethanol, dithiothreitol, or lipoic acid.

Temperature	The optimal temperature is 37 °C. *In vitro* cells cannot survive if the temperature increases beyond 40 °C, thus the temperature has to be regularly monitored. There are certain cell lines which are temperature sensitive for their phenotypic expresssion.	Temperature plays an important role in controlling the solubility of carbon dioxide, an increase in temperature increases solubility.
Osmolality	The ideal range of osmolality is 260–320 mosm kg^{-1}, which is in the range of human plasma osmolality, 290 mosm kg^{-1}.	An osmometer can be utilized to regularly monitor the osmolality of media, which can easily be affected by the addition of salts or drugs.

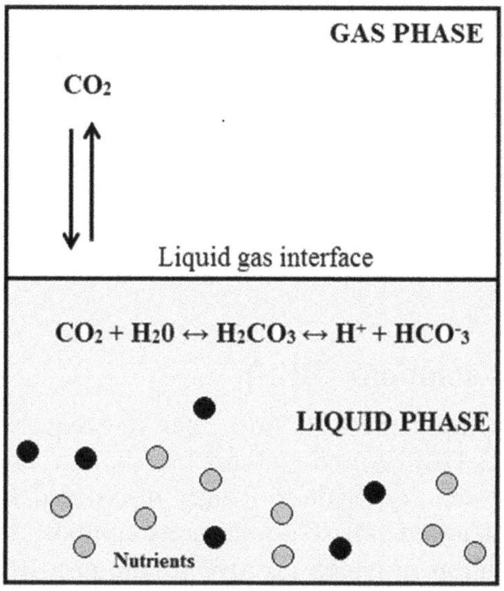

Figure 4.3. pH control by biocarbonate buffering.

century, many investigations relating cellular adaptation to experimental hypoxia have been carried out. A modular incubator chamber made of solid materials has frequently been used in experiments that require hypoxic conditions.

For this purpose, several models have been studied. One of the early facilities that was introduced was a modular incubator chamber that can be filled with low O_2 gas containing 1% O_2, 5% CO_2 and 94% N_2 [16–25]. The compartment is made of solid materials in a fixed shape and size. This type of facility has been extensively used as a hypoxia chamber in research laboratories in the past few decades. The major drawbacks of this facility are leakage and the creation of an inner pressure if the operation is inappropriate. Another model, supplied by an external high-pressure liquid nitrogen tank, is a cell culture incubator based on the displacement of O_2 with infusion of N_2 [16–25]. Another currently available facility is the hypoxia workstation, which offers a hypoxic environment for long-term cell cultures by controlling O_2 and CO_2 as well as temperature and relative humidity [16–20]. A hypoxic environment can also be developed by cells treated with cobalt chloride [16–25]. The various facilities available to maintain such conditions are listed in table 4.2.

Generally, most animal tissue culture experiments are performed under higher oxygen concentrations, where the cells are exposed to high oxidative stress, which adversely affects the cells and tissues in culture. Supplementation of some radical scavengers or antioxidants, as mentioned in table 4.1, helps in relieving cells of oxidative stress, in particular in the case where serum is not supplemented to the medium, as serum contains vital antioxidants. Additionally, certain organic salts and their respective ions, such as iron and copper ions in a free state, also encourage the generation of reactive oxygen species. Thus complexing agents have to be added with suitable carriers (e.g. transferrin, albumin or chelating agents). Care has to be taken with free iron ions, which can easily hydroxylate and precipitate in solution.

Table 4.2. Commercially available facilities of gas-tight/flush boxes.

Techniques	Purposes
Hypoxia/modular incubator chamber, flush box	Provide accurate, consistent hypoxic conditions
Hypoxia chamber for cell culture	Provide accurate, consistent hypoxic conditions
O_2 control cabinet for *in vitro* studies	Provide accurate, consistent hypoxic conditions

4.3 Balanced salt solutions (BSS)

BSSs mainly contain inorganic salts and offer the required pH and osmolality. Occasionally, glucose, HEPES and sodium bicarbonate are also added to BSSs. During the late 18th century, Sydney Ringer discovered Ringer's solution. This solution is called a balanced salt solution, as its composition is similar to that of body fluids. This solution provided positive results in a frog heart experiment in which the heartbeat was maintained even after dissection and removal from the body [26, 27]. This first breakthrough revolutionized the *in vitro* culture technology of animal tissue. Since then, various BSSs were developed, such as Locke's solution [28], Tyrode's solution [29], the Krebs–Ringer bicarbonate solution [30], Gey's solution [31], Earle's solution [32] and Hanks' solution [33]. The simple composition (inorganic salts, sometimes with glucose added as a nutrient) of BSS encourages its utilization. In addition, phenol red can be added to check pH variations. Several parameters, such as pH, osmotic pressure and inorganic salt concentration, of BSSs are usually standardized to meet the requirements of physiological conditions. BSSs can be utilized effectively to maintain tissues and cells under *in vitro* environments for short periods, usually up to a few days. The compositions of two BSSs are described in table 4.3.

4.4 Growth medium

Various nutrients such as amino acids, vitamins and serum are added to BSSs to encourage the growth of cells in culture. During the mid-20th century, Harry Eagle investigated the least amount of low-molecular-weight compounds required by mouse L cells and HeLa cell cultures for growth and proliferation. Following Fischer's method, he supplemented dialyzed serum in a BSS and realized that 13 amino acids and eight vitamins are essential for the growth and proliferation of mouse L cells and HeLa cells [34–36].

Later, he identified the minimal essential components, such as glucose, six inorganic salts, 13 amino acids, eight water-soluble vitamins and dialyzed serum, and developed the minimum essential medium (MEM) [37]. Based on the cell type and culture, the purpose of this medium was modified by Dulbecco and Vogt [38], Stanners *et al* [39], and Iscove and Melchers [40] (table 4.4). A major breakthrough in the field of cell medium development occurred when McCoy *et al* used dialyzed

Table 4.3. Ingredients of two BSSs.

BSS	Ingredients										
	NaCl	KCL	$CaCl_2$	$MgSO_4.7H_2O$	$NaHCO_3$	$NAH_3PO_4.H_2O$	$Na_2HPO_4.7H_2O$	KH_2PO_4	D-glucose	Phenol red	HEPES
Earle's BSS	6.68	0.4	0.02	0.2	2.2	0.14	–	–	1.0	0.001	13.02
Hank's BSS	8.0	0.4	0.14	0.1	0.35	–	0.09	0.06	1.0	0.001	2.08

Table 4.4. The development of different media.

Researcher	Year	Medium and composition	Advantages/applications	Ref.
Exclusion of serum and proteins				
White	1946	Developed a chemically defined medium that was composed of glucose, inorganic salts, amino acids, iron, vitamins and glutathione, with no protein at all.	The medium is devoid of serum and proteins. Developed chick embryo-derived fibroblasts and cardiac muscle cells for ~2 months.	[9, 45]
Parker	1950	Developed medium 199, which includes fat-soluble vitamins, cholesterol and nucleic acid precursors, in addition to the ingredients of White's medium.	As an alternative to serum or proteins, Parker's team used the strategy of adding low-molecular-weight substances essential for cell culture. By using this medium, they cultured chick embryo-derived cells for 3–4 weeks.	[46]
Serum-free culture for the best growth of mouse L cells				
Connaught Medical Research Laboratories (CMRL) media, Medium 199 (Morgan et al 1950)	1950	Developed in order to cultivate chicken embryonic cells under protein-free conditions, it is prepared by the sequential addition of amino acids, vitamins (including fat-soluble vitamins), and nucleic-acid precursors.	Used for organ culture	[34, 36]
Parker	1957–61	Developed a chemically defined medium, CMRL1066, consisting of 58 components, through several modifications, such as increasing the levels of reducing substances (cysteine, glutathione and ascorbic acid), removing fat-soluble vitamins, changing the nucleic acid precursors and adding co-enzymes.	Designed for the best growth of mouse L cells	[47, 48]

(*Continued*)

Evans	1959	Developed a chemically defined medium that was composed of 68 ingredients, named NCTC109. This media was difficult to prepare because of their complex composition.	Designed for the best growth of mouse L cells	[49]
Waymouth	1959	Developed the MB 752/1 medium, including 40 ingredients (glucose, inorganic salts, amino acids, vitamins, purine bases, hypoxanthin, and glutathione). This was the simplest possible chemically defined medium at the time.	Designed for the best growth of mouse L cells	[50]
Evolution of protein (serum fraction) media (cell cloning was not possible in protein-free media)				
Waymouth	1959	Demonstrated the auto/paracrine effect in which proteins that are synthesized by cells are essential for proliferation under protein-free conditions.	Designed for the best growth of mouse L cells	[50]
Ham	1963	Developed Ham's F-10 medium which contains two types of serum protein fractions (albumin and fetuin) instead of serum. This medium has an undefined composition because it contained serum protein fractions.	This was utilized to successfully make a single Chinese hamster ovary (CHO) cell form a colony under serum-free conditions.	[51]
Serum supplementation is essential for the cultivation of cell types				
Ham	1963	Developed a synthetic medium of definite composition, Ham's F-12, containing low-molecular-weight substances—linoleic acid and putrescine.	For serum-free single-cell plating of CHO cells.	[52]
Ham	1965	Developed MCDB301, a medium that is supplemented with 20 trace elements.	For cultivation of hybridomas.	[53–56]
Takaoka and Katsuta	1971	Successfully developed various kinds of long-term cultured cells in a simple medium that did not contain proteins or lipids; however, only a few cell types can adapt to protein- and lipid-free conditions. Thus cultured differs greatly from the cells cultured in serum-containing media.		[57, 58]

serum and identified pyruvate as an essential amino acid for the growth of carcinosarcoma cells [41].

During the same time, the USA-based Roswell Park Memorial Institute (RPMI) improved the calcium and magnesium concentrations in the 5A medium, discovered using McCoy's 5A medium, which is now called the RPMI 1640 medium [42–44]. The potential of this medium was observed later when it was first utilized to culture lymphocyte [42–44]. However, Eagle is still acknowledged as the most renowned discoverer of media, as his discovery led to the developments of several different media, as shown in table 4.4.

In addition to serum, a complete medium ideally contains amino acids, vitamins, salts, glucose, other organic supplements, growth factors, hormones and antibiotics.

4.4.1 The role of amino acids in animal culture

Amino acids are vital for the development of mammalian cells (figure 4.4). The significance of amino acids was realized early, particularly during the development of serum-free media, when a solution of a mixture of amino acids is supplemented in media for the development of cultured cells. The essential role of amino acids in terms of biological and chemical properties was first explored in chemically defined mammalian cell culture media. The concentrations of amino acids in media have traditionally been adjusted based on physiological requirements, in particular cellular consumption rates. However, due to biochemical conversion, it is challenging to evaluate amino acid requirements as it directly or indirectly affects amino acid consumption. The roles of several amino acid transporters (membrane transport proteins that transports amino acids) have been explored to evaluate the ultimate amino acid requirements of the cell. These proteins control the intracellular–extracellular exchange of these molecules. Such an exchange is easily influenced by conditions in cell culture media, thus better knowledge of the dissolution kinetics, stability, solubility and interactions of these molecules is required. This understanding helps in the utilization of these properties in the design and development of amino acid supplemented defined chemical media for mammalian cells. Biochemical conversion of amino acids involves several intermediate molecules which form metabolic networks. These networks can be analyzed using an approach called flux balance analysis (a mathematical procedure for simulating metabolism in

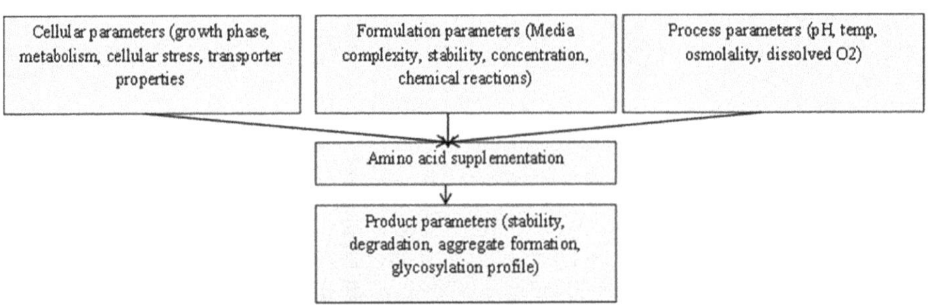

Figure 4.4. The role of amino acids in animal culture.

reconstructions of metabolic networks). Investigating amino acid concentrations, their rate of consumption from a medium and their metabolic rate helps in understanding how mammalian cells in culture interact with their environment [38, 59]. This will offer better understanding of the chemical behavior of these molecules in solutions of complex mixtures. The importance of amino acids was realized after the discoveries of Eagle (1955) and Dulbecco and Freeman (1959), who created nutrient supplements containing amino acids and vitamins that allowed for the cultivation of cells in adherent monolayers [38, 59]. Essential amino acids which cannot be produced by the cells must be supplemented to the medium, whereas non-essential amino acids which cannot be produced by the cells are also supplemented to meet the requirements of biosynthetic pathways or any metabolic processes [38, 59]. Glutamine and glutamate are considered as excellent sources of energy and carbon, thus they should be added in sufficient amounts to the medium [38, 59].

4.4.2 Vitamins

Cell culture media contain several components and each component has a defined role. The complete medium is a chemically well-defined basal medium supplemented with more or less defined additives [60]. One quality that increases the utility of the basal medium is that it contains low molecular weight substances such as inorganic ions, amino acids, vitamins and some other components (e.g. glucose, pyruvate, etc), however, basal medium and even enriched basal medium are not suitable for the growth of mammalian cells *in vitro* as they do not contain the necessary proteins for the suitable growth of mammalian cells [60].

Supplementation of high molecular weight substances in serum, in particular proteins, is required for the optimum growth of mammalian cells. In addition to these supplements media also contain peptides, lipoproteins, phospholipids or lipids [62]. Vitamin supplementation is essential, since the cells usually cannot synthesize vitamins, however, the quantity required is very small. Vitamins act as cofactors for many enzymes and deficiency can result in a decrease in cell growth, cell death or loss of productivity. The growth and development of the medium under *in vitro* conditions is also determined by the quantity and quality of vitamins, e.g. the water soluble vitamins present in Eagle's MEM medium and the fat soluble vitamins present in M 199. Media without serum require high concentrations of vitamins as serum itself contains certain vitamins.

4.4.3 Micro-nutrients

Salts are the main components of culture media, and result in the production of inorganic ions in aqueous solutions after their dissociation, e.g. embryo culture media contain six inorganic ions: sodium, potassium, chloride, calcium, magnesium, sulfate and phosphate ions. One of the most classical saline solutions is Krebs–Ringer bicarbonate. The composition of Whitten's medium, which was the first successful defined mouse embryo culture medium, was similar to Krebs–Ringer bicarbonate. A similar composition was used in the mouse embryo culture medium,

M16. In the early 1970s, human embryos were cultured in undefined somatic cell media supplemented with serum. After the development of a defined medium (Quinns HTF) in the 1980s for human embryo culture, several other media have been developed which have similar inorganic ion concentrations to M16 and Whitten's medium [61]. The advantage of such a medium is that it can be utilized for laboratory work and clinically.

Micronutrients such as minerals and vitamins also play an important role in culture media, and their supplementation is necessary for DNA metabolic pathways and therefore essential for life. Without their supplementation, the genomic stability of the cell line would be compromised, which may disturb homeostasis, resulting in chronic diseases and different types of malignant disorders. Cell culture media should offer a similar environment to that *in vivo*, and can be considered as *in vitro* models to assess cells responses against different stimuli. The supplementation of micronutrients allows a culture to better mimic the *in vivo* environment, which can ultimately increase cell viability and genomic stability. In many culture media, fetal bovine serum (FBS) is used as the sole source of micronutrients, which contribute only 5%–10% of the composition of the medium. In several studies, culture media supplemented with vitamins and minerals at concentrations not equivalent to physiological levels have been tested to evaluate the cell response. So far less attention has been paid to studying FBS composition, micronutrient assays during the cell cycle, the overall micronutrients in cell cultures, and the effect of micronutrients on the viability and genetics of cultured cells. Further investigations are also required to evaluate the roles of micronutrients at a molecular level and their effects on the genomic stability of cells.

Micronutrients are required for optimal macronutrient metabolism because of their critical role in intermediate metabolism. Invariably, metabolism requires the concomitant involvement of one or more vitamins and minerals. Chronic degenerative disease etiology and the rate of pathogenesis are thus intimately associated with micronutrient imbalances. Nutrition research has recently highlighted the role of several nutrients in regulating the genomic machinery [90–93]. More specifically, a number of vitamins and micronutrients are substrates and/or cofactors in the metabolic pathways regulating DNA synthesis and/or repair and gene expression [90–93]. A deficiency in such nutrients may result in the disruption of genomic integrity and alteration of DNA methylation, thus linking nutrition with the modulation of gene expression. In many cases, the response to a nutrient deficiency also seems to be genotype specific. Gene-nutrient interactions are thus a fascinating example of physiological responses to the environment/diet at the molecular level [90–93].

As discussed above micronutrients are important to DNA metabolic pathways, however, there is no report available that supports the statement that these micronutrients optimally protect against DNA damage. Several studies are available related to *in vitro* model demonstration, which includes the roles of micronutrients in maintaining genomic stability (table 4.5).

Table 4.5. Roles of different micronutrients.

Micronutrients	Roles	References
Vitamins C	Prevents DNA oxidation and chromosomal damage	[94, 95]
Vitamin E	Prevents DNA oxidation and chromosomal damage	[96]
Vitamin D	Antioxidant activity, stabilizes chromosomal structure, and prevents DNA double-strand breaks	[96]
Magnesium	Essential cofactor in DNA metabolism that plays a role in maintaining the high fidelity of DNA transcription	[97]
Iron	Excess of or a deficiency in iron may cause DNA breaks	[98]
Carotenoid	Carotenoid-rich diet reduces DNA damage	[99]
Retinol	Excess retinol may be carcinogenic	[100]
Vitamin B12	Vitamin B12 deficiency is associated with the formation of micronuclei. Vitamin B12 is required for growth, genetic stability and survival of cells in vitro. It supports one-carbon metabolism and the degradation of amino and odd-chain fatty acids, respectively.	[90–93]
Transcobalamin II	Reduced transcobalamin II in the serum is associated with chromosomal abnormalities. Transcobalamins I and II as natural transport proteins of vitamin B12. Transcobalamin II appears to have a physiological function in plasma transport of B12. Cobalamins are water soluble molecules that do not cross cell membranes efficiently. They generally enter cells as a complex with transcobalamin II	[101]

Table 4.6. Cellular functions of various ions.

Ions	Cellular functions
Calcium ions	• Adhesion • Signal transduction • Cell proliferation and differentiation
Sodium, potassium and chloride ions	• Regulating membrane potential
Phosphate, sulfate and bicarbonate ions	• Maintaining intracellular charge • Act as precursors for the production of certain compounds such as ATP

Given the importance of micronutrients *in vitro*, the optimization of cell viability and genomic stability warrants further study. Cell culture media mimicking the *in vivo* environment may help to generate *in vitro* models of a cell's response to different stimuli. The composition of these media includes certain vitamins and minerals, but unfortunately, in many common culture media, the only source of micronutrients is FBS, which makes up only 5%–10% of the medium. Moreover, an appropriate proportion of micronutrients is not always provided because the precise composition of each batch of FBS is in fact extremely variable [34].

The salt composition of several media is quite similar to the salt composition present in BSSs. These salts offer unique ions (i.e. cations and anions) which help in maintaining osmolality and can be part of a cofactor, essential for protein synthesis [61]. Some of the major functions or various ions are listed in table 4.6.

4.4.4 Glucose

The glucose added to animal tissue culture media, once consumed by mammalian cells, is converted in lactate. The rate of glucose consumption is influenced by glucose concentration. During batch processes the concentration of glucose can be manipulated by programmed supplementation of glucose. In batch cultures the specific feeding rate of glucose and the proportion of glucose converted to lactate can be reduced. A decreased rate of conversion of glucose to lactate seems to overlap with an increased oxygen uptake rate [62]. One of the possibilities for such programmed feeding of glucose is the increased oxidation of glutamine and the simultaneous augmented production of ammonium. High amounts of ammonium and lactate can be growth inhibitory for the cultivation of hybridoma cells [62]. Glucose acts as an instantaneous source of energy. During glycolysis, glucose is converted into pyruvate/lactase. Pyruvate/lactase enters into mitochondria in order to be fully oxidized by the citric acid cycle and into CO_2. It has been reported that compared to glucose, glutamine provides more carbon under *in vitro* conditions, although glucose is still considered as an active source of carbon under *in vivo* conditions [62].

Glucose and glutamine metabolism have been tested under *in vitro* conditions in several cultured mammalian cell lines (BHK, CHO and hybridoma cell lines) by correlating specific utilization and formation rates with specific maximum activities of the regulatory enzymes involved in glycolysis and glutaminolysis. A high degree of similarity in many aspects of glucose and glutamine metabolism was observed among the cultured mammalian cell lines examined [63]. The effect of glucose concentration in the medium on rat hepatocyte culture has been recently investigated and it was observed that low-glucose culture medium is better than high-glucose culture medium for culturing of hepatocytes [64]. The effect of disaccharides as an energy source in protein-free mammalian cell cultures has been evaluated recently. Mammalian cells are usually unable to utilize polysaccharides for cell growth because the phospholipid bilayer in the cell membrane has very low permeability to sugars. It was discovered that mammalian cells can utilize maltose for growth in the absence of glucose and successfully adapt and grow in glucose-free, maltose-containing, serum- and protein-free media [65].

4.4.5 Hormones and growth factors

During the 1960s insulin was initially utilized as a supplement/growth factor for culture media [66, 67]. In the beginning the independent use of insulin was not considered as effective as serum [67, 68]. Later, it was realized that the utilization of insulin in combination with low-concentration serum offers a higher level of efficacy for baby hamster kidney (BHK)21 cell growth. This breakthrough has encouraged other researchers to utilize insulin with an optimum level of serum. From the mid-20th century, various reports have been published that hastened the development of serum-free media. Generally, hormones and growth factors are not required in media supplemented with serum, however, they are essential in serum-free media. Several growth factors were explored, such as nerve growth factor [69], epidermal growth factor [70, 71], insulin-like growth factor [72–75], fibroblast growth factor (FGF) [76, 77], platelet-derived growth factor [77–79] and transforming growth factor (TGF) [80, 81]. Supplementation of these growth factors to a culture medium resulted in improved cellular proliferation. However, in comparison to serum, their actual effect on cellular growth was found to be inferior in practice [82–88].

4.4.6 Additional organic supplements

Organic supplements such as biological proteins, peptides, lipids, nucleosides and various citric acid cycle intermediates are used to support growth, in particular in serum-free media [89].

4.4.6.1 Antibiotics

Previously, the utilization of antibiotics in culture media was very common. Ampicillin, penicillin, gentamycin, neomycin, kanamycin, erythromycin and tetracycline were often utilized to reduce the risks of contamination. However, with the advancements in aseptic technologies, currently the supplementation of antibiotics is not required. In fact, supplementation of antibiotics can result in the following disadvantages:

- The possible development of antibiotic resistant cells in cultures.
- Inhibited biological metabolic processes, which can ultimately result in under-development of cells.
- Infections can be masked for a period of time.
- Promotes a poor aseptic environment.

Currently antibiotic use is not recommended, however, they can be utilized for the development of primary cultures, because the transition from *in vivo* to *in vitro* environments can result in exposing the explant to various contaminants.

According to the standard cell culture guidelines, supplementation of antibiotics is ideally required to prevent cell contamination, however, comparatively little information is available about the effects of antibiotic use in cell cultures on gene expression; antibiotic treatment could result in confused results. Recently, it has been suggested that penicillin–streptomycin treatment can considerably affect gene expression and regulation in liver cells, which supports that antibiotic treatment should be taken into consideration when carrying out genetic, genomic or other biological assays in cultured cells [102].

The effects of gentamicin on cellular physiology were also investigated on several mammalian cell lines and it has been proved that gentamicin is superior for the control of bacterial growth in tissue culture [103, 104].

The advancements in knowledge related to the use of antibiotics during culture should include the effects of antibiotics over the cellular differentiation potential. The antibiotics not only alter the genetic expression but also cell biochemistry and may modify the differentiation potential of cultured cells. Recently, human adipose tissue-derived stem cells were differentiated to adipocytes with or without antibiotics commonly used in these culture protocols, such as a penicillin–streptomycin–amphotericin mix or gentamicin, and the results were evaluated. It has been shown that these antibiotics affect cell differentiation. Effects on melanogenic efficiency have also been evidenced after the utilization of penicillin and streptomycin, which can affect the melanogenic activity of malignant melanocytes in culture [105, 106].

It was found that of the antibiotics, penicillin–streptomycin, which is one of the most commonly used antibiotics in cell culture media to control bacterial contamination, and gentamicin, widely used for the treatment of infections caused by gram-negative bacteria, did not affect human ES cell viability or the expression of pluripotency markers [107–109]. However, when differentiation was directed towards neural and hepatic fates, substantial cell death was noted due to the activation of the caspase cascade. In addition, the expression of neural progenitor markers was considerably reduced, signifying that gentamicin may negatively affect early embryonic neurogenesis. The treatment of animals with high doses of gentamicin shows necrosis of proximal kidney tubular cells [110, 111], whereas low doses of gentamicin induced programmed cell death through the activation of the caspase cascade [111]. Moreover, it has been observed that therapeutic doses of gentamicin cause hearing loss and nephrotoxicity in neonates [112, 113]. Although it is known that aminoglycosides can cross the placenta, the effect of maternal use of these antibiotics on early embryonic development, if any, is still not well known.

4.5 Serum

In a laboratory setting, cell culture technology is one of the most common procedures used to study human disease environments. Cell culture technologies allow the growth and maintenance of cells of various types, including cells obtained from primary tissues, such as stem cells and cancer tumors. However, a central confusing issue in cell cultures is the utilization of serum and animal products in media, which can lead to batch and lot variations that ultimately result in experimental variability, confuses studies with therapeutic outcomes for cultured cells, and signifies a main cost related to cell culture technology [114]. Some of the commercially available serum-free, albumin-free and xeno-free media (e.g. NeuroPure™) are more cost-effective than other commercial media [114]. Serum is a nutrient enriched natural biological fluid which supports cell growth and development. The addition of basal culture media with animal serum of diverse origin is important for cell growth and for the stimulation of proliferation. This is called the mitogenic effect. The most extensively used serums are derived from adult or newborn animals, such as FBS, which usually contains a mixture of a large number of constituents. These constituents are low and high molecular weight substances with different, physiologically balanced growth-promoting and growth-inhibiting activities. Certain substitutes for serum such as protein hydrolysates have also been studied in cell culture medium formulations [115]. Recently, vegetable hydrolysates have also been used as serum substitutes to support the amino acid content in small peptide form for batch and fed-batch fermentations [115]. These novel hydrolysates, such as peptones of soy, rice, wheat gluten, etc, have been recently examined as protein-free medium supplements for the production of a recombinant therapeutic protein. It has been discovered that peptones offer nutritional benefits to recombinant cell lines [115]. The well-known FBS is a universal growth supplement of cell and tissue culture media, as it is a natural mixture of several nutritional components which are required for cell attachment, growth and proliferation, and is effective for most types of human and animal (including insect) cells.

Although in use for more than 50 years, FBS has never been fully characterized. Recent proteomic and metabolomic studies revealed approximately 1800 proteins [116, 117] and more than 4000 metabolites [118] present in the serum. However, the use of serum in cell culture also has a number of disadvantages. These disadvantages can be seen from the following perspective: (a) cell biology, since serum in general is an ill-defined mixture of components in culture media, with qualitative and quantitative, geographical and seasonal batch-to-batch variations; (b) biosafety, since FBS may contain adverse factors, such endotoxins, mycoplasma, viral contaminants or prion proteins; (c) ethics, in terms of animal protection arguments regarding the harvest and collection of FBS from bovine fetuses; and (d) recent concerns about the global supply versus demand of FBS. As a consequence, a number of strategies have been developed to reduce or replace the need for FBS in cell culture media.

Suitable combinations of medium, cells and serum are listed in table 4.7. Calf serum, horse serum and fetal serum are the commonest examples of serum. Before using these serums they must be screened for viral pathogens. The optimal concentration for its supplementation is 5%–20% which varies from medium to medium. Some of the essential components of serum are briefly described in the following sections.

4.5.1 Proteins

Albumin is the most abundant protein in serum and is multifunctional. It binds to various water-insoluble substances such as lipids. The *in vitro* roles of these proteins are not yet known, however, some proteins encourage cell attachment and growth. In addition to their buffering action, most proteins increase the viscosity of media because of their high molecular weight.

4.5.2 Nutrients and metabolites

Serum is full of nutrients such as amino acids, glucose, phospholipids, fatty acids, nucleosides and metabolic intermediates. Serum provides proteins, nutrients,

Table 4.7. A list of suitable combinations of medium, cell line and serum.

Medium	Cell or cell line	Serum
RPMI 1640	Human leukemia	FB
RPMI 1640	Mouse leukemia	FB, HoS
RPMI 1640	Mammary epithelium	FB
RPMI 1640	Hematopoietic cells	FB
MEM	3T3 cells	CS
EMEM	Chick embryo fibroblasts	CS
EMEM	Chinese hamster ovary	CS
EMEM	HeLa cells	CS
DMEM, MEM	Glial cells	FB
DMEM	Neurons	FB
DMEM	Skeletal muscle	FB

EMEM: Eagle's minimum essential medium; RPMI: Roswell Park Memorial Institute medium; DMEM: Dulbecco's modified Eagle's medium; MEM: minimum essential medium; CS: calf serum; FB: fetal bovine serum; HoS: horse medium.

attachment factors, trace elements, growth factors, and hormones, which aids in the growth of the cells. Although FBS is the most commonly used serum product, many other products are sold as a lower cost alternative.

4.5.3 Growth factors

Growth factors such as platelet derived growth factor, fibroblast growth factor, epidermal growth factor and insulin-like growth factor present in media always promote cell proliferation. Most of the growth factors in serum are now commercially available. These growth factors always act in small amounts on cells, e.g. to trigger proliferation, differentiation, migration, secretion or importation. Under serum-free conditions, several cells types need growth factor supplementation in the medium. In the absence of heparin sulfate (over the surface of the target cells) acidic fibroblast growth factor breakdown occurs, thus heparin is always supplemented to avoid its degradation in the medium. However, growth factor supplementation has some considerable limitations, for example animal-derived growth factors can cause contamination and certain growth factors can inhibit the growth of epithelial cells, e.g. transforming growth factor-β. Some other growth factors are epidermal growth factor, fibroblast growth factor, insulin-like growth factor, Madin-Darby canine kidney, nerve growth factor, platelet-derived growth factor and transforming growth factor. It is now understood that for culture purposes, natural clot serum is more effective than plasma in stimulating cell proliferation. This is because of the release of certain polypeptides from activated platelets during the clotting process. Recently, several growth factors, hormones, transport proteins, cofactors and essential minerals and trace elements have been identified, that all drive specific gene expression, initiate and control the cell cycle and thus cell division, and program specific cell differentiation.

4.5.4 Hormones

Hormones such as hydrocortisone, triiodothyronine, estrogen, androgens, progesterone, prolactin, follicle-stimulating hormone and gastrin-releasing peptide are utilized for their specific roles, e.g. hydrocortisone encourages cellular attachment while insulin allows glucose uptake by cells. Growth hormones, in combination with somatomedins, encourage cell proliferation. Similarly, growth hormone in association with insulin encourages the proliferation of a variety of cells. In addition to encouraging cellular attachment, hydrocortisone increases the cloning efficiency of glial cells and fibroblasts. This is required for the maintenance of epidermal keratinocytes and several other endothelial cell types. Another good example of a hormone is triiodothyronine, supplementation of which is required for Madin-Darby canine kidney epithelial cells and is used for pulmonary epithelium. In association with hydrocortisone and prolactin, different blends of estrogen, androgens and progesterone are required for the maintenance of the mammary epithelium. Hormones also have certain considerable disadvantages, such as the issue of stability, e.g. a high concentration of cysteine always negatively affects the stability of insulin. Similarly, for optimal action of insulin, zinc supplementation is required. In a closely packed culture, for example a culture of pulmonary epithelial cells, hydrocortisone supplementation offers growth inhibition, whereas in low-density cultures it occasionally encourages growth.

4.5.5 Carrier proteins

Carrier proteins such as transferrin, albumin, lactoferrin and others are utilized as carrier proteins for different substances as follows:

- Amino acids such as cysteine and tryptophan.
- Lipids such as fatty acids and cholesterol.
- Trace elements such as copper and nickel.
- Vitamins such as vitamin B6.

Because of their poor solubility in aqueous solutions, these carrier proteins form complexes with lipids to effectively deliver a suitable amount of lipids to cells. Features such as its anti-toxin, antioxidant and anti-stress effects makes albumin more favorable for utilization. For transporting iron, the carrier protein transferrin is recommended, and lactoferrin can also serve as alternative to transferrin. These carrier proteins sometimes increase the chance of contamination, such as viral contamination, in particular when they are derived from serum. In most cases, the serum-derived albumin available today is purified from corn through a cold ethanol fractionation procedure. The advantage of this approach is that the protein derived by this method does not contain other proteins, however, the amount of lipids and trace elements bound to the albumin differs from batch to batch. Thus, researchers should execute prior batch screening before utilizing these products. As serum-derived transferrins are derived from three different sources, such as porcine, bovine and human, there is variation in the activity, e.g. bovine serum-derived transferrin is not active at lower concentrations, thus its concentration should be elevated to obtain significant results.

4.5.6 Lipids and related components

Several lipid and lipid-like compounds are utilized, such as cholesterol, steroids, fatty acids, ethanolamine, choline and inositol. Lipids form biological membranes (phospholipid membranes), store and transport nutrients and also play an important role in signal transduction. Cell lines under *in vitro* conditions can synthesize the lipids required for metabolism from acetyl coenzyme A, but supplementing the lipids can decrease the biosynthetic load. Moreover, certain cells do not have the desirable enzymes required for cholesterol biosynthesis, in such circumstances sterols must be supplemented to the medium. However, lipid supplementation also has considerable disadvantages, such as the need for ethanol, surfactants and cyclodextrin to solubilize the lipids in circumstances where the culture medium should be used under protein-free conditions. The amount of ethanol required for solubilization can have considerable negative effects on cells, thus precautions have to be taken before its utilization. Similarly, the utilization of surfactants and cyclodextrin is also criticized as these compounds can cause toxicity to cultured cells at high concentrations.

4.5.7 Transition metals

Different transition metals, such as iron, zinc, copper, chromium, iodine, cobalt, manganese and molybdenum, have been utilized for their property of experiencing electron transfer. Due to this property, they become part of the active centers of enzymes and other bio-macromolecules present inside the cell. Selenium, iron, copper and zinc are usually used in cell cultures. Selenium, when present in the form of seleno-proteins, exhibits antioxidant properties, such as thioredoxin reductase. Iron can act as a chelating agent, however, any variation in supplementing the optimal quantity of iron can lead to the production of reactive oxygen species.

4.5.8 Vitamins

Vitamins are essential for cell division, growth and development. Additionally, they also act as precursors of various cofactors. Broadly, vitamins are divided into two classes A, D, E and K (the fat-soluble vitamins) and B1, B2, B6, B12, C and folate (the water-soluble vitamins). For antioxidant effects vitamins C and E are often utilized. The type of vitamin used, in particular fat-soluble vitamins, is often decided by the cell type, and vitamins are usually present in most basal media. Oxidation caused by air can degrade vitamins A, C, D and E. Additionally vitamin C degradation often occurs due to oxidation by trace elements, which may sometimes lead to the production of free radicals. Certain vitamins, such as vitamins A, B1, B2, B12, C and K, are degraded by light and for some (vitamins B1 and B5) heat is a cause of degradation. As far as solubility is concerned, folate exhibits poor solubility and is partially removed after filtration sterilization. Using hydroxocobalamin and vitamin C in the same medium encourages their mutual degradation. Hence, physicochemical parameters should be regularly monitored to prevent the degradation of vitamins.

4.5.9 Polyamines

Polyamines, such as putrescine, spermidine and spermine, are the low-molecular-weight basic physiologically active substances that exist universally in cells and encourage protein or nucleic-acid synthesis. The concentration of polyamines are regulated and maintained by intracellular biosynthesis or decomposition and transport from outside the cell. A decrease in polyamine concentration in the cell stops the growth of the cell. This decrease in concentration can be due to disturbances in polyamine biosynthesis, decomposition and transport. In addition, a sudden increase in polyamine synthesis can also result in apoptosis.

4.5.10 Reducing agents

Several reducing agents, such as 2-mercaptoethanol, α-thioglycerol and reduced glutathione, are used for multiple purposes. In microbiology these reducing agents are utilized to develop and maintain anaerobic conditions for culturing anaerobic micro-organisms. In animal tissue culture the purpose of the supplementation of cysteine in media is to maintain the intracellular redox environment, in particular when the cells lack the cystine transporters that convert cystine into cysteine. However, we should be careful when supplementing any reductants as their supplementation in the presence of albumin can cause severe damage to cells.

4.5.11 Additives

Certain additives, such as detergents, polyvinyl pyrrolidone, carboxymethyl cellulose, F-68, tween 80, Pluronic and others, are supplemented to decreases the shear stress offered by stirred cultures. Certain solubilizers in the form of surfactants are utilized to solubilize lipophilic substances (e.g. lipids, fat-soluble vitamins), however, certain surfactants can cause cytotoxicity at higher concentrations. Stabilizing and detoxifying factors are required to maintain pH or to inhibit proteases either directly, e.g. antitrypsin or macroglobulin, or indirectly, by acting as an unspecific sink for proteases and other (toxic) molecules.

4.5.12 Adhesion factors

Adhesion factors such as fibronectin and laminin help to anchor organized tissue cells to the interface. These factors are required if the cells are anchorage-dependent. The supplementation of adhesion factors in media always encourages the adhesion of anchorage-dependent cells to vessels, however, sometimes it may also increase the chances of viral contamination in the case where the components for support are derived from viral sources.

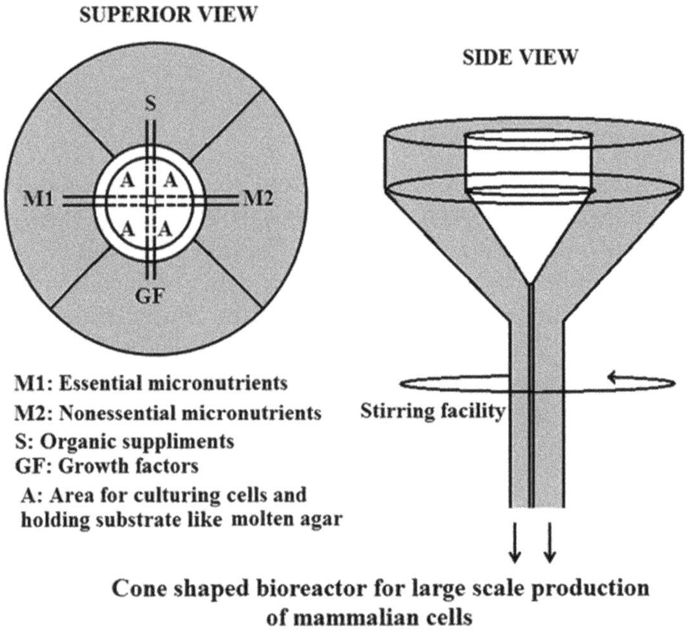

Figure 4.5. A hypothetical bioreactor design.

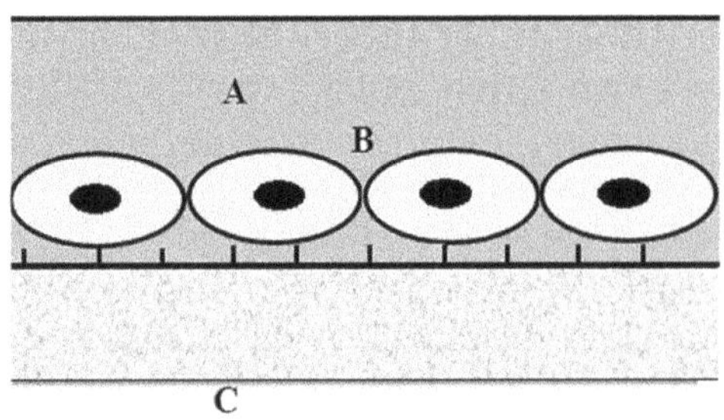

Figure 4.6. The *in vivo* and *in vitro* microenvironment of the mammalian cell.

4.6 Bioreactors in animal tissue culture

Different designs of bioreactors are available that allow the rapid and uniform mixing of all ingredients, including macro- and micronutrients, natural supplements and growth factors (figure 4.5). A hypothetical version of a new cone shaped bioreactor is shown in this figure which not only allows uniform mixing but also offers ease in harvesting desirable cultured cells. The nutrients present in the four different compartments are allowed to mix in the central area where controlled conditions are set for optimum culture of animal cells. The flow rate of nutrients towards the center chamber should be maintained in such a way that it will not restrict cell growth. Suitable 3D structures/scaffolds can also be utilized at the center to meet the requirements of the culture.

The cellular microenvironment is presented schematically in figure 4.6. The microenvironment contains a culture substratum in the form of a Petri dish or specially coated surfaces, or perhaps a feeder layer that encourages attachment of cells over the surface. The substratum also encourages cell spreading, which encourages cell–cell adhesion. The environment offered by the culture medium allows the diffusion of nutrients from the medium to cells/micro-organisms. These nutrients contain salts, hormones and growth factors. As discussed in this chapter, the culture medium must contain all the essential nutrients required for cell metabolism, growth and proliferation. Animal tissue culture medium must contain:

- Biosynthetic precursors for cellular anabolism.
- Catabolic substrates for energy metabolism.
- Vitamins and trace elements, primarily for catalytic activity.
- Inorganic ions in surplus quantities for catalytic activity and to maintain physiological conditions.
- Animal serum, for chemically defined basal media.

The different aspects that regulate the behavior of cells in culture are: (a) the diffuse environment offered by the culture medium; (b) the terminal connections between adjacent cells, i.e. cell–cell adhesion; and (c) the contact environment, i.e. cell–matrix adhesion.

4.7 Serum-free media

Serum-free media do not require the supplementation of serum for optimal cell growth, however, they may contain other additional components derived from animals, such as lactalbumin, casein, insulin, lipids and sterols [120]. Some serum-free media are categorized as animal-component-free media (ACFM) in which none of the components are animal-derived [119]. Protein-free media do not contain polypeptide factors but may contain hydrolyzed peptide fragments from animal or plant sources. Chemically defined media (CDM) contain defined components of low molecular weight and are, in most cases, free of proteins [119, 120]. There are certain cells which have already been investigated for use in serum-free media, such as BHK21 cells which are well adapted to grow in serum-free or animal-component-free cell media for rabies vaccine production [121, 122]. In their adaptation to serum-free conditions, BHK21 cells showed a switch from anchorage-dependent to suspension growth [121, 122] and certain basic changes in cell structure [123, 124].

The traditional approach of supplementing serum in media is still utilized to obtain a reasonable level of growth. Serum-free media have been developed for the following reasons:

- To avoid any chances of contamination.
- To define medium composition.
- As it is difficult to maintain the uniform quality of the serum batch to batch.
- To avoid the presence of growth inhibitors.
- For continuous supply (serum derived from cattle is dependent on the availability of the cattle).

Table 4.8. Composition of serum-free media.

Growth factors	Hormones	Nutrients	Proteins	Polyamines	Protease inhibitors
These growth factors work synergistically with other growth factors or other factors such as hormones or prostaglandins: epidermal growth factor (EGF), fibroblast growth factor (FGF), platelet derived growth factor (PDGF), nerve growth factor (NGF), insulin-like growth factor (IGF) cytokines (interferons, interleukins), erythropoietin (EPO), eye-derived growth factors (EDGF 1, EDGF 2), interleukins (IL1, IL2), hepatocyte growth factor (HGF), lipopolysaccharides (LPS), phytohemagglutinin (PHA)	insulin, hydrocortisone, dexamethasone, trijodo-thyronine, thyroxine, progesterone, estradiol, prolactin, glucagon, vasopressin, PTH. A combination of hydrocortisone, estrogen, androgen and progesterone is utilized for the maintenance of mammary epithelium.	choline, ethanolamine, linoleic acid, iron, copper, selenium, etc	bovine serum, albumin	Putrescine, cadaverine, spermine, spermidine, β-phenylethylamine	α1-antitrypsin, serine, protease inhibitor
		Trace elements sodium and selenite	**Fatty acids and lipids** linoleic acid, oleic acid cholesterol, phospholipids ethanolamine	**vitamin A-derivatives**, retinoic acid, retinol-acetate	**Adhesion factors** coating of culture vessels with collagen (type I, type IV), fibronectin, laminin

Table 4.9. Cell lines with suitable serum-free media.

Cell line	Medium
Prostatic epithelium	WAJC 404
Mammary epithelium	MCDB 170
Human vascular endothelium	MCDB 131
Human lung fibroblasts	MCDB 110
Fibroblasts	MCDB 202
Chinese hamster ovary	MCDB 402
Chick embryo fibroblasts	MCDB 202
Bronchial epithelium	LHC 9
3T3 cells	MCDB 402

- The presence of serum interferes with or disturbs the cell harvesting process, which can ultimately affect the entire down-stream process.

Serum-free media offer all the advantages which are not offered by serum media. One of the main advantages is to control the growth of cells by offering a defined medium, in contrast to serum containing media where cells grow in an uncontrolled fashion. Moreover, serum-free media offer a factor, or set of growth factors, that helps in regulating the differentiation of cells. However, despite these benefits, serum-free media offer considerable disadvantages, such as:

- A slow cell proliferation rate.
- Depending upon the nature of the cell line, the composition and type of medium also varies, thus multiple media are required for different cell lines, and in certain cases different media are required at different development stages for the same cell line, which creates many difficulties.
- The requirement for standard chemical reagents (pure chemicals) for the preparation of the medium.
- The use of pure chemicals makes them more expensive then serum containing media.
- May require growth factors and hormones which are difficult to isolate and purify.
- Serum free media cannot be developed for all types of cell lines.
- Require optimization for each hybridoma cell line.

4.7.1 Preparation of serum-free media

Serum-free media are normally used for more cell-specific cultures. Before the preparation of a serum-free medium it is essential to identify the components of serum and their respective concentrations [125, 126]. The main components of natural serum are:

- Growth regulatory factors such as TGF-β.
- Cell adhesion factors such as vitamins.

- Essential nutrients such as minerals, metabolites, fatty acids and vitamins.
- Hormones such as insulin and hydrocortisone.
- Adhesion (attachment and spreading) factors.

The components of serum-free media are listed in table 4.8, and suitable combinations of cell lines and serum-free media are listed in table 4.9.

References

[1] Yao T and Asayama Y 2017 Animal-cell culture media: history, characteristics, and current issues *Reprod. Med. Biol.* **16** 99–117

[2] Jacquez J A and Barry E 1951 Tissue culture media : the essential non-dialyzable factors in human placental cord serum *J. Gen. Physiol.* **34** 765–74

[3] Merten O W 2006 Introduction to animal cell culture technology—past, present and future *Cytotechnology* **50** 1–7

[4] Lodish H *et al* 2000 Growth of animal cells in culture *Molecular Cell Biology* 4th edn (New York: Freeman) section 6.2

[5] Orr H C, Baker J and Cheesman J O 1973 Survival of animal tissue cells in primary culture in the absence of serum *Appl. Microbiol.* **25** 49–54

[6] Healy G M and Parker R C 1966 Cultivation of mammalian cells in defined media with protein and nonprotein supplements *J. Cell Biol.* **30** 539–53

[7] Morton H J, Pasieka A E and Morgan J F 1956 The nutrition of animal tissues cultivated *in vitro* : III. use of a depletion technique for determining SPECIFIC nutritional requirements *J. Biophys. Biochem. Cytol.* **2** 589–96

[8] Hegde A and Behr B 2012 Media composition: growth factors *Methods Mol. Biol.* **912** 177–98

[9] Jacoby F and Darke S J 1948 Animal tissue culture with a synthetic medium *Nature* **161** 768–9

[10] Okamoto T, Sato J D, Barnes D W and Sato G H 2013 Biomedical advances from tissue culture *Cytotechnology* **65** 967–71

[11] Eagle H 1971 Buffer combinations for mammalian cell culture *Science* **174** 500–3

[12] Poole C A, Reilly H C and Flint M H 1982 The adverse effects of HEPES, TES, and BES zwitterion buffers on the ultrastructure of cultured chick embryo epiphyseal chondrocytes *In Vitro Cell. Dev. Biol.* **18** 755–65

[13] Zigler J S Jr, Lepe-Zuniga J L, Vistica B and Gery I 1985 Analysis of the cytotoxic effects of light-exposed HEPES-containing culture medium *In Vitro Cell. Dev. Biol.* **21** 282–7

[14] Jang J, moon S J, Hong S H and Kim I H 2010 Colorimetric pH measurement of animal cell culture media *Biotechnol. Lett.* **32** 1599–607

[15] Grady L H, Nonneman D J, Rottinghaus G E and Welshons W V 1991 pH-dependent cytotoxicity of contaminants of phenol red for MCF-7 breast cancer cells *Endocrinology* **129** 3321–30

[16] Walsh-Reitz M M and Toback F G 1992 Phenol red inhibits growth of renal epithelial cells *Am. J. Physiol. Renal. Physiol.* **262** F687–91

[17] Wang R, Jin F and Zhong H 2014 A novel experimental hypoxia chamber for cell culture *Am. J. Cancer Res.* **4** 53–60

[18] Semenza G L 2012 Hypoxia-inducible factors in physiology and medicine *Cell* **148** 399–408

[19] Wang G L, Jiang B H, Rue E A and Semenza G L 1995 Hypoxia-inducible factor 1 is a basic-helix-loop-helix-PAS heterodimer regulated by cellular O_2 tension *Proc. Natl Acad. Sci. USA* **92** 5510–4
[20] Jaakkola P *et al* 2001 Targeting of HIF-alpha to the von Hippel–Lindau ubiquitylation complex by O_2-regulated prolyl hydroxylation *Science* **292** 468–72
[21] Safran M and Kaelin W G Jr 2003 HIF hydroxylation and the mammalian oxygen-sensing pathway *J. Clin. Invest.* **111** 779–83
[22] Post D E and Van Meir E G 2003 A novel hypoxia-inducible factor (HIF) activated oncolytic adenovirus for cancer therapy *Oncogene* **22** 2065–72
[23] Edin N J, Olsen D R, Sandvik J A, Malinen E and Pettersen E O 2012 Low dose hyper-radiosensitivity is eliminated during exposure to cycling hypoxia but returns after reoxygenation *Int. J. Radiat. Biol.* **88** 311–9
[24] Esteban M A and Maxwell P H 2005 Manipulation of oxygen tensions for *in vitro* cell culture using a hypoxic workstation *Expert Rev. Proteomics* **2** 307–14
[25] Rodgers G M, Fisher J W and George W J 1975 The role of renal adenosine 3',5'-monophosphate in the control of erythropoietin production *Am. J. Med.* **58** 31–8
[26] Ringer S 1882 Concerning the influence exerted by each of the constituents of the blood on the contraction of the ventricle *J. Physiol.* **3** 380–93
[27] Ringer S 1883 A further contribution regarding the influence of the different constituents of the blood on the contraction of the heart *J. Physiol.* **4** 29–42
[28] Locke F S and Rosenheim O 1907 Contributions to the physiology of the isolated heart: the consumption of dextrose by mammalian cardiac muscle *J. Physiol.* **36** 205–20
[29] Tyrode M 1910 The mode of action of some purgative salts *Arch. Int. Pharmacodyn.* **20** 205–23
[30] Krebs H and Henseleit K 1932 Untersuchungen über die Harnstoff Bildung im Tier Körper *Klin. Wochenschr.* **11** 757–9
[31] Gey G O and Gey M K 1936 The maintenance of human normal cells and tumor cells in continuous culture: I. Preliminary report: cultivation of mesoblastic tumors and normal tissue and notes on methods of cultivation *Am. J. Cancer* **27** 45–76
[32] Earle W, Schilling E, Stark T, Straus N, Brown M and Shelton E 1943 Production of malignancy *in vitro*; IV: the mouse fibroblast cultures and changes seen in the living cells *J. Natl Cancer Inst.* **4** 165–212
[33] Hanks J H and Wallace R E 1949 Relation of oxygen and temperature in the preservation of tissues by refrigeration *Proceedings of the Society for Experimental Biology and Medicine* vol 71 (New York: Society for Experimental Biology and Medicine) pp 196–200
[34] Eagle H 1955 The specific amino acid requirements of a mammalian cell (strain L) in tissue culture *J. Biol. Chem.* **214** 839–52
[35] Eagle H 1955 The specific amino acid requirements of a human carcinoma cell (strain HeLa) in tissue culture *J. Exp. Med.* **102** 37–48
[36] Eagle H 1955 Nutrition needs of mammalian cells in tissue culture *Science* **122** 501–14
[37] Eagle H 1959 Amino acid metabolism in mammalian cell cultures *Science* **130** 432–7
[38] Dulbecco R and Freeman G 1959 Plaque production by the polyoma virus *Virology* **8** 396–7
[39] Stanners C P, Eliceiri G L and Green H 1971 Two types of ribosome in mouse–hamster hybrid cells *Nature New Biol.* **230** 52–4

[40] Iscove N N and Melchers F 1978 Complete replacement of serum by albumin, transferrin, and soybean lipid in cultures of lipopolysaccharide-reactive b lymphocytes *J. Exp. Med.* **147** 923–33

[41] Neuman R E and McCoy T A 1958 Growth-promoting properties of pyruvate, oxaloacetate, and α-ketoglutarate for isolated Walker carcinosarcoma 256 cells *Exp. Biol. Med.* **98** 303–6

[42] McCoy T A, Maxwell M and Kruse P F 1959 The amino acid requirements of the Jensen sarcoma *in vitro* *Cancer Res.* **19** 591–5

[43] Moore G E, Ito E, Ulrich K and Sandberg A A 1966 Culture of human leukemia cells *Cancer* **19** 713–23

[44] Moore G E, Gerner R E and Franklin H A 1967 Culture of normal human leukocytes *J. Am. Med. Assoc.* **199** 519–24

[45] White P R 1946 Cultivation of animal tissues *in vitro* in nutrients of precisely known constitution *Growth* **10** 231–89

[46] Morgan J F, Morton H J and Parker R C 1950 Nutrition of animal cells in tissue culture. I. Initial studies on a synthetic medium *Exp. Biol. Med.* **73** 1–8

[47] Parker R C, Castor L N and McCulloch E A 1957 Altered cell strains in continuous culture. A general survey *Spec. Publ. NY Acad. Sci.* **5** 303–13

[48] Parker R C 1961 *Chemically Defined Media: Methods of Tissue Culture* (New York: Harper Row) pp 62–80

[49] Evans V J *et al* 1956 Studies of nutrient media for tissue cells *in vitro* II. An improved protein-free chemically defined medium for long-term cultivation of strain L-929 cells *Cancer Res.* **16** 87–94

[50] Waymouth C 1959 Rapid proliferation of sublines of NCTC clone 929 (strain L) mouse cells in a simple chemically defined medium (MB 752/1) *J. Natl Cancer Inst.* **22** 1003–17

[51] Ham R G 1963 Albumin replacement by fatty acids in clonal growth of mammalian cells *Science* **140** 802–3

[52] Ham R G 1963 Putrescine and related amines as growth factors for a mammalian cell line *Biochem. Biophys. Res. Commun.* **14** 34–8

[53] Ham R G 1965 Clonal growth of mammalian cells in a chemically defined, synthetic medium *Proc. Natl Acad. Sci. USA* **53** 288–93

[54] Hayashi I and Sato G H 1976 Replacement of serum by hormones permits growth of cells in a defined medium *Nature* **259** 132–4

[55] Murakami H 1989 Serum-free media used for cultivation of hybridomas *Adv. Biotechnol. Process.* **11** 107–41

[56] Hamilton W G and Ham R G 1977 Clonal growth of Chinese hamster cell lines in protein-free media *In Vitro Cell. Dev. Biol.* **13** 537–47

[57] Takaoka T and Katsuta H 1971 Long-term cultivation of mammalian cell strains in protein- and lipid-free chemically defined synthetic media *Exp. Cell Res.* **67** 295–304

[58] Oshima M, Miyamoto I, Takaoka T, Goto K and Kagawa Y 1987 Unsaturated fatty acids in mammalian cell strains cultured for more than 10 years in serum-free synthetic media *Jpn J. Exp. Med.* **57** 351–4

[59] Salazar A, Keusgen M and von Hagen J 2016 Amino acids in the cultivation of mammalian cells *Amino Acids* **48** 1161–71

[60] Büntemeyer H and Lehmann J 2001 The role of vitamins in cell culture media *ESACT Proc. Animal Cell Technology: From Target to Market* ed E Lindner-Olsson, N Chatzissavidou and E Lüllau vol 1 (Dordrecht: Springer)

[61] Baltz J M 2012 Media composition: salts and osmolality *Methods Mol. Biol.* **912** 61–80
[62] Hu W S, Dodge T C, Frame K K and Himes V B 1987 Effect of glucose on the cultivation of mammalian cells *Dev. Biol. Stand.* **66** 279–90
[63] Neermann J and Wagner R 1996 Comparative analysis of glucose and glutamine metabolism in transformed mammalian cell lines, insect and primary liver cells *J. Cell Physiol.* **166** 152–69
[64] Na G H, Kim D G, Kim Y H, Han J H and Jung E S 2014 Effects of glucose concentration in the medium on rat hepatocyte culture *Ann. Surg. Treat. Res.* **87** 53–60
[65] Leong D S Z, Tan J G L, Chin C L, Mak S Y, Ho Y S and Ng S K 2017 Evaluation and use of disaccharides as energy source in protein-free mammalian cell cultures *Sci. Rep.* **7** 45216
[66] Lieberman I and Ove P 1959 Growth factors for mammalian cells in culture *J. Biol. Chem.* **234** 2754–8
[67] Higuchi K 1963 Studies on the nutrition and metabolism of animal cells in serum-free media: I. Serum-free monolayer cultures *J. Infect. Dis.* **112** 213–20
[68] Clarke G D, Stoker M G P, Ludlow A and Thornton M 1970 Requirement of serum for DNA synthesis in BHK 21 cells: effects of density, suspension and virus transformation *Nature* **227** 798–801
[69] Clarke G D and Stoker M G P 1971 Conditions affecting the response of cultured cells to serum *Ciba Foundation Symp. Growth Control in Cell Cultures* ed G E W Wolstenholme and J Knight (Chichester: Wiley) pp 17–32
[70] Levi-Montalcini R 1952 Effects of mouse tumor transplantation on the nervous system *Ann. NY Acad. Sci.* **55** 330–44
[71] Cohen S 1962 Isolation of a mouse submaxillary gland protein accelerating incisor eruption and eyelid opening in the new-born animal *J. Biol. Chem.* **237** 1555–62
[72] Taylor J M, Mitchell W M and Cohen S 1972 Epidermal growth factor: physical and chemical properties *J. Biol. Chem.* **247** 5928–34
[73] Salmon W D Jr and Daughaday W H 1957 A hormonally controlled serum factor which stimulates sulfate incorporation by cartilage *in vitro J. Lab. Clin. Med.* **49** 825–36
[74] Froesch E R, Buergi H, Ramseier E B, Bally P and Labhart A 1963 Antibody-suppressible and nonsuppressible insulin-like activities in human serum and their physiologic significance. An insulin assay with adipose tissue of increased precision and specificity *J. Clin. Invest.* **42** 1816–34
[75] Daughaday W H, Hall K, Raben M S, Salmon W D Jr, van den Brande J L and van Wyk J J 1972 Somatomedin: proposed designation for sulphation factor *Nature* **235** 107
[76] Rinderknecht E and Humbel R E 1976 Polypeptides with nonsuppressible insulin-like and cell-growth promoting activities in human serum: Isolation, chemical characterization, and some biological properties of forms I and II *Proc. Natl Acad. Sci. USA* **73** 2365–9
[77] Gospodarowicz D 1974 Localisation of a fibroblast growth factor and its effect alone and with hydrocortisone on 3t3 cell growth *Nature* **249** 123–7
[78] Gospodarowicz D 1975 Purification of a fibroblast growth factor from bovine pituitary *J. Biol. Chem.* **250** 2515–20
[79] Kohler N and Lipton A 1974 Platelets as a source of fibroblast growth-promoting activity *Exp. Cell Res.* **87** 297–301 .
[80] Ross R, Glomset J, Kariya B and Harker L 1974 A platelet-dependent serum factor that stimulates the proliferation of arterial smooth muscle cells *in vitro Proc. Natl Acad. Sci. USA* **71** 1207–10

[81] Antoniades H N, Scher C D and Stiles C D 1979 Purification of human platelet-derived growth factor *Proc. Natl Acad. Sci. USA* **76** 1809–13
[82] De Larco J E, Preston Y A and Todaro G J 1981 Properties of a sarcoma-growth-factor-like peptide from cells transformed by a temperature-sensitive sarcoma virus *J. Cell Physiol.* **109** 143–52
[83] Anzano M A *et al* 1982 Communication: synergistic interaction of two classes of transforming growth factors from murine sarcoma cells *Cancer Res.* **42** 4776–8
[84] Hollenberg M D and Cuatrecasas P 1973 Epidermal growth factor: receptors in human fibroblasts and modulation of action by cholera toxin *Proc. Natl Acad. Sci. USA* **70** 2964–8
[85] Gospodarowicz D and Moran J S 1974 Stimulation of division of sparse and confluent 3t3 cell populations by a fibroblast growth factor, dexamethasone, and insulin *Proc. Natl Acad. Sci. USA* **71** 4584–8
[86] Holley R W and Kiernan J A 1974 Control of the initiation of DNA synthesis in 3t3 cells: serum factors *Proc. Natl Acad. Sci. USA* **71** 2908–11
[87] Smith G L and Temin H M 1974 Purified multiplication-stimulating activity from rat liver cell conditioned medium: comparison of biological activities with calf serum, insulin, and somatomedin *J. Cell Physiol.* **84** 181–92
[88] Gospodarowicz D and Moran J 1975 Optimal conditions for the study of growth control in BALB/ 3t3 fibroblasts *Exp. Cell Res.* **90** 279–84
[89] Fenech M and Ferguson L R 2001 Vitamins/minerals and genomic stability in humans *Mutat. Res.* **475** 1–6
[90] Ames B N 2001 DNA damage from micronutrient deficiencies is likely to be a major cause of cancer *Mutat. Res.* **475** 7–20
[91] Fenech M 2002 Micronutrients and genomic stability: a new paradigm for recommended dietary allowances (RDAs) *Food Chem. Toxicol.* **40** 1113–7
[92] Ames B N 1999 Micronutrient deficiencies. A major cause of DNA damage *Ann. NY Acad. Sci.* **889** 87–106
[93] Arigony A L *et al* 2013 The influence of micronutrients in cell culture: a reflection on viability and genomic stability *Biomed. Res. Int.* **2013** 597282
[94] Halliwell B 2001 Vitamin C and genomic stability *Mutat. Res.* **475** 29–35
[95] Claycombe K J and Meydani S N 2001 Vitamin E and genome stability *Mutat. Res.* **475** 37–44
[96] Chatterjee M 2001 Vitamin D and genomic stability *Mutat. Res.* **475** 69–88
[97] Hartwig A 2001 Role of magnesium in genomic stability *Mutat. Res.* **475** 113–21
[98] De Freitas J M and Meneghini R 2001 Iron and its sensitive balance in the cell *Mutat. Res.* **475** 153–9
[99] Collins A R 2001 Carotenoids and genomic stability *Mutat Res.* **475** 21–8
[100] Wang Z *et al* 2012 Dietary intakes of retinol, carotenes, vitamin C, and vitamin E and colorectal cancer risk: the Fukuoka colorectal cancer study *Nutr. Cancer* **64** 798–805
[101] Rana S R, Colman N, Goh K O, Herbert V and Klemperer M R 1983 Transcobalamin II deficiency associated with unusual bone marrow findings and chromosomal abnormalities *Am. J. Hematol.* **14** 89–96
[102] Ryu A H, Eckalbar W L, Kreimer A, Yosef N and Ahituv N 2017 Use antibiotics in cell culture with caution: genome-wide identification of antibiotic-induced changes in gene expression and regulation *Sci. Rep.* **7** 7533

[103] Fischer A B 1975 Gentamicin as a bactericidal antibiotic in tissue culture *Med. Microbiol. Immunol.* **161** 23–39
[104] Perlman D 1979 Use of antibiotics in cell culture media *Methods Enzymol.* **58** 110–6
[105] Llobet L, Montoya J, López-Gallardo E and Ruiz-Pesini E 2015 Side effects of culture media antibiotics on cell differentiation *Tissue Eng.* C **21** 1143–7
[106] Martínez-Liarte J H, Solano F and Lozano J A 1995 Effect of penicillin–streptomycin and other antibiotics on melanogenic parameters in cultured B16/F10 melanoma cells *Pigment Cell Res.* **8** 83–8
[107] Fischer A B 1975 Gentamicin as a bactericidal antibiotic in tissue culture *Med. Microbiol. Immun.* **161** 23–39
[108] Quiros Y, Vicente-Vicente L, Morales A I, López-Novoa J M and López-Hernández F J 2011 An integrative overview on the mechanisms underlying the renal tubular cytotoxicity of gentamicin *Toxicol. Sci.* **119** 245–56
[109] Guthrie O W 2008 Aminoglycoside induced ototoxicity *Toxicology* **249** 91–6
[110] Parker R A, Bennett W H and Porter G A 1982 Animal models in the study of aminoglycoside nephrotoxicity *The Aminoglycosides: Microbiology, Clinical Use and Toxicology* ed A Whelton and H C Neu (New York: Dekker) pp 235–67
[111] El Mouedden M, Laurent G, Mingeot-Leclercq M P and Tulkens P M 2000 Gentamicin-induced apoptosis in renal cell lines and embryonic rat fibroblasts *Toxicol. Sci.* **56** 229–39
[112] Musiime G M, Seale A C, Moxon S G and Lawn J E 2015 Risk of gentamicin toxicity in neonates treated for possible severe bacterial infection in low- and middle-income countries: systematic review *Trop. Med. Int. Health* **20** 1593–606
[113] Cross C P, Liao S, Urdang Z D, Srikanth P, Garinis A C and Steyger P S 2015 Effect of sepsis and systemic inflammatory response syndrome on neonatal hearing screening outcomes following gentamicin exposure *Int. J. Pediatr. Otorhinolaryngol.* **79** 1915–9
[114] Usta S N, Scharer C D, Xu J, Frey T K and Nash R J 2014 Chemically defined serum-free and xeno-free media for multiple cell lineages *Ann. Transl. Med.* **2** 97
[115] Heidemann R *et al* 2000 The use of peptones as medium additives for the production of a recombinant therapeutic protein in high density perfusion cultures of mammalian cells *Cytotechnology* **32** 157–67
[116] Aerson N L and Anderson N G 2002 The human plasma proteome. History, character, and diagnostic prospects *Mol. Cell Proteomics* **1** 845–67
[117] Anderson N *et al* 2004 The human plasma proteome: a nonredundant list developed by combination of four separate sources *Mol. Cell Proteomics* **3** 311–26
[118] Psychogios N *et al* 2011 The human serum metabolome *PLoS ONE* **65** e16957
[119] Merten O W 2002 Development of serum-free media for cell growth and production of viruses/viral vaccines—safety issues of animal products used in serum-free media *Dev. Biol.* **111** 233–57
[120] Jayme D W and Smith S R 2000 Media formulation options and manufacturing process controls to safeguard against introduction of animal origin contaminants in animal cell culture *Cytotechnology* **33** 27–36
[121] Kallel H, Jouini A, Majoul S and Rourou S 2002 Evaluation of various serum and animal protein free media for the production of a veterinary rabies vaccine in BHK-21 cells *J. Biotechnol.* **95** 195–204

[122] Rourou S, Ben Ayed Y, Trabelsi K, Majoul S and Kallel H 2014 An animal component free medium that promotes the growth of various animal cell lines for the production of viral vaccines *Vaccine* **32** 2767–9

[123] Amadori M and Berneri C 1993 Genotypic and phenotypic changes of BHK-21 cells grown in suspension cultures *Cytotechnology* **11** S106–8

[124] Amadori M, Volpe G, Defilippi P and Berneri C 1997 Phenotypic features of BHK-21 cells used for production of foot-and-mouth disease vaccine *Biologicals* **25** 65–73

[125] Bhatia S 2018 *Introduction to Pharmaceutical Biotechnology: Basic Techniques and Concepts* vol 1 (Bristol: IOP Publishing)

[126] Bhatia S 2018 *Introduction to Pharmaceutical Biotechnology: Enzymes, Proteins and Bioinformatics* vol 2 (Bristol: IOP Publishing)

IOP Publishing

Introduction to Pharmaceutical Biotechnology, Volume 3
Animal tissue culture and biopharmaceuticals
Saurabh Bhatia, Tanveer Naved and Satish Sardana

Chapter 5

Animal tissue culture facilities

5.1 Introduction

When planning animal tissue culture facilities, utmost care should be taken to maintain aseptic conditions, which is perhaps one of the most under-appreciated aspects of tissue culture. Ideally, animal cell culture studies should be performed in a single-use facility which should be separated into a zone for handling newly received material, called the quarantine area, and an area for material which is known to be free of contaminants, referred to as the main tissue culture facility. All new material should be handled as 'quarantine material' until it has been shown to be free of contaminants such as bacteria, fungi and particularly *Mycoplasma*. Good planning of the laboratory and its layout to meet the requirements of planned operations is a major challenge, however, the time invested at the initial phase can pay off with increased operational efficiency and effectiveness. The key elements of successful laboratory design are mentioned in figure 5.1.

The most important components with regard to infrastructure and equipment are listed in the following sections.

5.2 Infrastructural requirements

- Mechanical/plumbing/engineering: Dedicated or shared back-up generators; alarms and controlled access; CO_2 and liquid N_2 delivery to the laboratory and cryobank.
- A heating, ventilation and air conditioning (HVAC) system, for both comfort in the laboratory and for the biological safety cabinets which play a major role in the construction of a tissue culture laboratory.
- Interior finishes: vinyl flooring; nonporous ceilings; washable, impermeable paint and coatings; impermeable bench-tops and furniture.

doi:10.1088/2053-2563/aafac0ch5

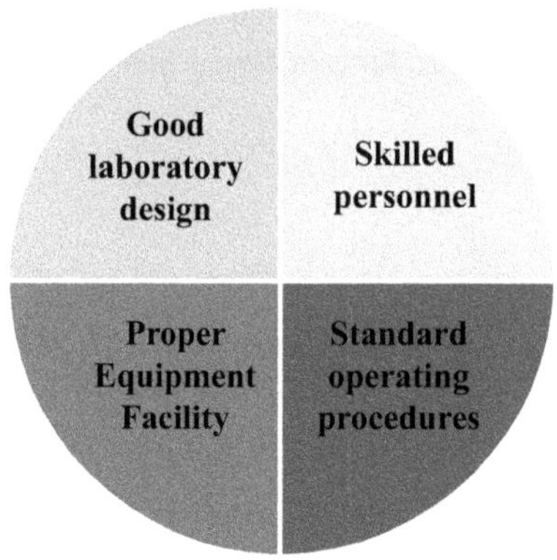

Figure 5.1. Successful animal tissue culture laboratory design has four key elements.

5.3 Equipment

The required equipment includes: a laminar flow (horizontal or vertical) device; moist heat sterilizer; refrigerator and freezer (−20 °C); digital balance; CO_2 cylinder; water purifier (Millipore water or double distillation assembly); liquid nitrogen; humidifier; air sterilizer; gas plasma sterilizer; ethylene oxide and formaldehyde based low temperature sterilizer; cryopreservation facility; and many others, listed in table 5.1.

5.4 Culture vessels

The tissue culture process involves the attachment of cells to the surface of a vessel which acts as a support in the form of a substrate. Therefore, depending upon the type of cell that is considered for culture, nature of the material changes. Some cells are anchorage-dependent cells, requiring attachment for their growth, whereas others are anchorage-independent and do not require any support for their growth. Anchorage-dependent cells can be detached from the culture vessel by a process called trypsinization, which uses the proteolytic enzyme trypsin to detach adherent cells from the surface of a cell culture vessel. Due to the recent advancements in cell culture technology, there is an immense demand for more reliable three-dimensional culture technology as two-dimensional technologies having considerable drawbacks, as mentioned in figure 5.2. There is immense scope for the development of diverse substrates to maintain a three-dimensional niche under an *in vitro* environment.

Certain cells have the tendency to form cell aggregates and to communicate with each other through cellular signaling and form mutual contacts. This helps to form specific microenvironments that enable them to express a tissue-like phenotype. One of the best examples of establishing microenvironments is cellular spheroids, which represent the most common use of *ex vivo* three-dimensional cultures. Cellular spheroids are three-dimensional models that can be produced from an extensive variety of cell types and form due to the tendency of adherent cells to aggregate. Some of the advancements in the field of three-dimensional culture technology are shown in figure 5.3.

Table 5.1. List of the equipment required to establish a standard animal tissue culture laboratory.

Sr. No.	Facility	Equipment's
1	Tissue culture laboratory	• Class II biosafety cabinet (BSC). • CO_2 incubator. • Pipettors. • Vacuum flask/aspiration device. • Water bath (37 °C). • Low-speed centrifuge (clinical grade, for spinning cells).
2	Microscopy	• Phase contrast microscope. • Dissecting microscope.
3	Storage	• Cabinets and shelves for the storage of tissue culture supplies. • Refrigerator (4 °C). • Freezer (−20 °C, nondefrosting). • Low temperature freezer (−70 °C to −85 °C). • Cryogenic freezer (storage below −140 °C, usually liquid nitrogen).
4	Molecular biology laboratory/quality control laboratory	• Reverse transcription polymerase chain reaction (RT-PCR). • Flow cytometer (might be in a core facility). • Fluorescence microscope (might be located in a microscopy core). • Confocal microscope (might be located in a microscopy core).

5	Quarantine laboratory	• Class II biosafety cabinet.
		• Incubators:
		  • Water or air jacketed.
		  • One gas (CO_2 and air).
		  • Two gases (CO_2, N_2 and air).
		  • Three gasses (CO_2, N_2 and O_2).
		• Phase contrast microscope.
		• Water bath (37 °C).
		• Low-speed centrifuge (clinical grade, for spinning cells).
		• Pipettors.
		• Aspiration/vacuum flask.
		• Sink.

6	Additional access to common equipment or core facilities	• Microscopy. • Flow cytometry. • Microarray gene expression. • Genomics. • Proteomics. • Virus production. • Vivarium.

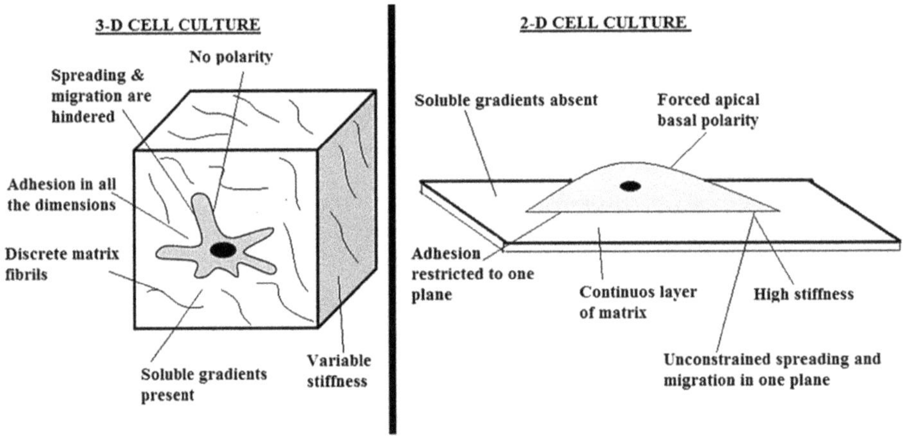

Figure 5.2. Three-dimensional cultures versus two-dimensional cultures.

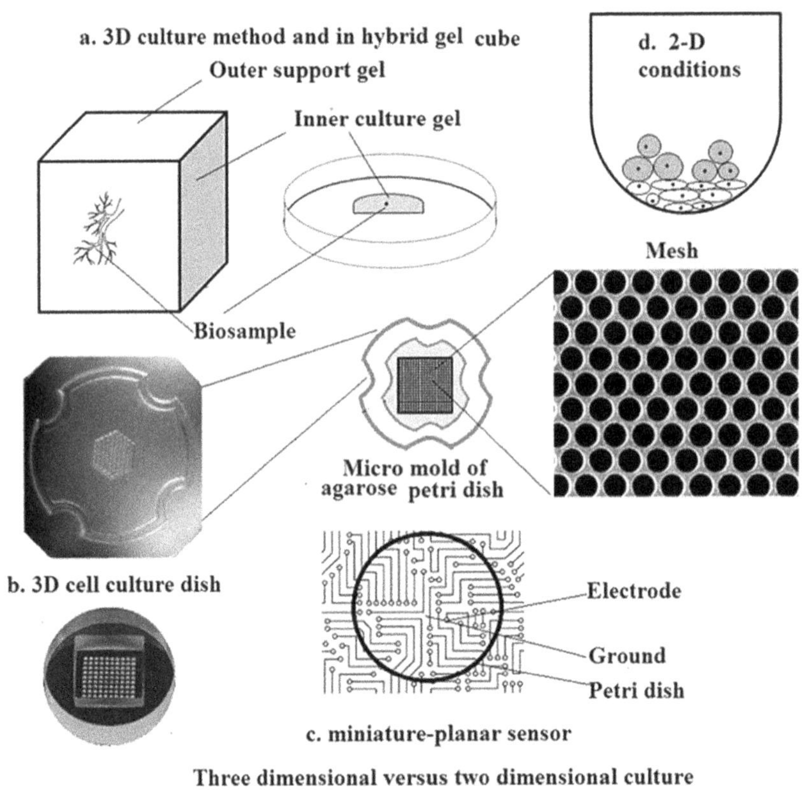

Figure 5.3. Advancements in the field of three-dimensional culture technology.

5.4.1 Materials used for culture vessels

Previously, disposable plastics and glass were used for culturing animal cells under adherent conditions. Because of the availability of more suitable substrates, the utilization of glass has almost been abandoned. Synthetic plastic materials such as polyvinyl chloride, polycarbonate, metinex and thermonex offer uniform and reproducible cultures because of their good consistency and optical properties. However, these materials still offer considerable limitations, such as:

- Limited surface area for cell growth.
- A nonhomogeneous culture environment.
- Challenges in monitoring.
- Challenges in controlling culture parameters (e.g. pH, dissolved oxygen).

Revolutionary developments in the field of anchorage-dependent culture allowed the development of human fibroblast-like cell cultures using solid microparticles/microcarriers. These microparticles/microcarriers were suspended in the culture medium by slow stirring [1–5]. This type of substrate contains beads of the ion exchange gel diethylaminoethyl-sephadex consisting of cross-linked dextran. This support system offers a positively charged culture surface and more suitable characteristics such as:

- Good cell adhesion properties.
- A large surface area/volume ratio.
- Good optical properties for microscopic visualization.
- Appropriate density.

This support system was successfully studied for mammalian cells [6, 7] and later used for the production of inactivated poliomyelitis vaccine [8]. Since the high charge of matrix was interfering with growth, a low-charged version of these microcarriers was later developed in the form of cytodex (dextran) microcarriers. Many factors, such as polarity and charge density, have been investigated to encourage cell adhesion, however, studies of extracellular matrix proteins, such as collagens, laminin and fibronectin, have resulted in drastic changes in the growth patterns of cells [9]. A novel type of gelatin microcarrier has been developed, containing a porous structure with an increased internal surface that allows cell growth inside microcarriers [10]. This discovery has resulted in the development of other matrices such as polystyrene, glass and cellulose [9–12], and various shapes such as cylindrical microcarriers [12]. These support systems were tested for the growth of different cell types and have been used to produce various vaccines and recombinant proteins, such as vaccines against poliomyelitis, rabies and foot-and-mouth disease, and recombinant proteins such as interferon and monoclonal antibodies [13–18].

5.4.2 Types of culture vessel

Some of the most commonly used culture vessels in animal tissue culture science are:

- Flasks.
- Multiwell plates.

- Petri dishes.
- Stirred bottles.

The choice of culture vessel depends on various factors such as:
- Cell type (anchorage-dependent monolayer or anchorage-independent suspension cells).
- Total quantity of cells required at the end of the process.
- Cost and purpose for which cells are cultured.
- Sampling frequency.

Generally the monolayer or anchorage-dependent cell yield is dependent on the surface area offered by the culture vessel. Flasks of diverse shapes and sizes (25–500 cm^2) are utilized for this purpose. These flasks save preparation time and avoid the risk of contamination. For specific purposes, and also to achieve optimum growth, speciality cell culture flasks for low adherence and spheroid cultures and coated flasks are also used. An extra wide flask neck offers easier access to the growth surface with scraper or pipette. Poly-D-lysine (PDL) and collagen I coated flasks are ideal for adherent cultures of primary cells that are finicky and have difficulty attaching to the growth surface. The uniform coating present in these flasks builds a positive charge over the surface that ensures cell attachment, growth and differentiation. These types of flasks can be used when cell lines are difficult to culture and show low adherence or slowed growth even after changing culture conditions. Fully molded well plates called solid plates made up of transparent polystyrene, or in black or white for use in fluorescence or luminescence studies, are also used. This material is designed in such a way that it offers low autofluorescence, the absence of a halo effect, improved optical clarity and high signal-to-noise ratios. These plates have excellent imaging properties and thus can be used for magnification up to 40×. For anchorage-dependent cells, polystyrene tubes are utilized to ensure consistent cell growth. Additionally, matrix screw top tubes to firmly store samples at low temperatures, at vapor phase (liquid nitrogen), can be utilized. Due to their high pore density (which allows more exchange through the membrane), polycarbonate cell culture inserts (multiwell plates) are currently being utilized for diverse applications, as shown in figure 5.4.

Additionally, chamber slides are designed for growth, fixation, staining and microscopic examination of cultured cells on a single surface with a removable chamber. Chambered cover glasses, which are made of thin glass and utilized for live cell imaging, are also utilized. Chambered slides are made of normal slide glass and are not designed to be imaged through. Instead, you remove the chamber unit after performing immunofluorescence staining in the wells and seal a cover glass over the top of the mounting medium.

5.4.3 Cell culture vessel treatment

To improve cell attachment and growth, a seeding procedure is used. Cell seeding is essential for cell proliferation, differentiation and extracellular matrix (ECM)

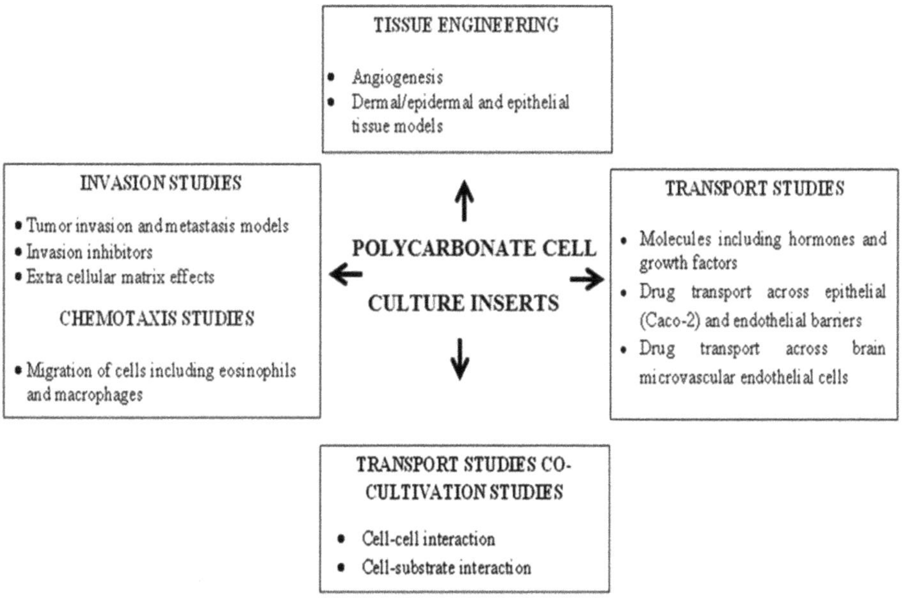

Figure 5.4. Applications of polycarbonate cell culture inserts.

synthesis [25, 26]. Earlier works proved that that an increased cell density increases biosynthesis of ECM such as collagen [20, 21]. Similarly, an increase in seeding density results in an increase in Alkaline Phosphatase synthesis and more mineralization [22]. Further work determined the ideal seeding density *in vitro* in a collagen gel41 and *in vivo* in a goat model [23]. Changing the cell seeding density was used as an approach to alter the paracrine signal distance, and thus to enhance the cell–biomaterial microenvironments and improve osteogenic signal expression [24].

Surface treatment of the substrate with fibronectin or collagen is required to increase the efficiency of cultured cells. To achieve this, the desired chemical (fibronectin or collagen) is poured over the surface of the substrate followed by the drying of the chemical and its sterilization under UV light. Additionally, the substrate's surface can also be treated with a monolayer of special cells which is referred to as the feeder layer.

The nature of the surface determines the cell's ability to attach, proliferate, migrate and function. ECM proteins are frequently used to coat glass or plastic surfaces to improve cell attachment *in vitro* [27]. ECM molecules can be immobilized over surfaces so as to imitate the effects seen by whole molecules. In culture vessels it has been observed that matrix coating the glass with purified ECM proteins such as collagen, laminin and fibronectin results in increased cell growth [27].

Pre-prepared matrix coated containers are now available as Matrigel, ProNectin and Cell-Tak, however, a coating is yet to be discovered which can 100% mimic the *in vivo* environment.

The growth of some tissues suddenly increases when they are cultured over a feeder layer, which supports the tissues through the release of end/metabolic products produced by living cells,, such as mouse embryo fibroblasts. Certain released products of these fibroblast cells can be used to increase the growth of other cells.

Alternative approaches, such as microcarriers and metallic substrates, can also be utilized for the propagation of cells under *in vitro* conditions. As mentioned above, the bead-shaped microcarriers are made up of collagen, gelatin, polyacrylamide and polystyrene. These substrates are used for anchorage-dependent cells in suspension. In addition to microcarriers, metallic substrates can also be utilized for certain cells. These cells can be conveniently cultured over metallic surfaces such as stainless steel disks or on palladium. A major reason for the use of metallic surfaces for culturing cells is their electrical conductivity. This property of electrically conductive substrates allows the cells to grow more efficiently and thus they can be used as cell-stimulating interfaces. Cellular responses to electrical conductivity have been studied extensively [28–32].

However, it has been observed that platinum substrates encouraged astrocyte cell growth, whereas gold surfaces exerted the opposite effect. This is because even some metallic surfaces exert too low a metallic conductivity to trigger cell growth [33].

5.4.4 Non-adhesive substrates

In certain cases cell adhesion is undesirable. In contrast to adhesive substrates such as plastic, glass, palladium, metallic surfaces, etc, non-adhesive substrates including agar, agarose and methyl cellulose (Methocel) can also be utilized. The first two are gels and the latter is a highly viscous solution where cells will sediment slowly. In this case cells are immobilized in the matrix and sustain growth. One of the criticisms related to the lack of growth of tumor cells over soft agar medium was addressed recently [34]. It was observed that the consistency of agar plays an important role in determining the growth of cells. Additionally, non-adhesive surfaces, such as ‘bacterial plastic’ dishes, are often sterilized with ethylene oxide. This treatment can impede the growth of the cells, so radiation sterilization must be utilized for culturing transformed cell lines [34].

5.5 Sterilization procedures in animal tissue cultures

Micro-organisms are present on all nonliving objects, creating a ubiquitous environment for potential contamination in the laboratory. In animal tissue culture, the success of experiments depends on the ability of a researcher to achieve sterilization and thus to eventually eliminate all micro-organisms. Additionally, the contact of sterile materials with non-sterile surfaces should be prevented. Contamination is a major concern in animal tissue culture, but certain methods can be adopted to avoid or eliminate contaminants [34, 35]. Bacteria and fungi (including molds and yeasts) are present everywhere in nature. These micro-organisms can grow over the surfaces of inanimate objects and also develop their colonies and multiply in rich cell culture environments. Most of our blood cells fall in the range of 100–300 μm, however, bacteria are small, 0.5–20 μm, so they can easily infect a culture and multiply within a host cell, thus making microbes the most commonly encountered cell culture contaminants [34, 35].

5.5.1 Biological contamination

Contamination is a widespread concern in the culturing of cells. Usually, for the prevention of contamination antibiotics can be utilized for short periods, however, regular utilization can result in the development of resistant strains and also masks any low-level infection. The accidental discovery of penicillin in 1928 by Scottish bacteriologist Alexander Fleming led to one of the greatest developments of modern medicine—the antibiotic era. After returning from holiday Fleming noticed that a Petri dish containing a staphylococcus culture which was left on a lab bench had been contaminated with a mold. After careful examination under a microscope, he detected that the mold had inhibited bacterial growth and created a zone of inhibition (a germ-free zone) in which the bacteria did not grow. He concluded that the mold must be synthesizing an antibacterial agent. This antimicrobial agent not only inhibited the growth of bacteria, but could be utilized to fight infectious diseases. The name of the mold was *Penicillium notatum* [36, 37]. This positive example of cell culture contamination is an exception, as contamination is a major problem in basic research and is the most difficult challenge for researchers developing and maintaining cultures. The study of different contaminants and their identification has resulted in the development of sterilization technologies and protocols [38]. In the mid-1990s, 11%–15% of cultures in US laboratories were infected with *Mycoplasma* species [39]. Even after taking all the necessary measures, the prevalence rate of mycoplasma infection progressed up to 23%. The prevalence rate was, surprisingly 8.45% in 2010 for biopharmaceuticals, where most of the cultures were contaminated with fungi and bacteria, including *Mycoplasma* [40]. Contamination is considered one of the most important uncontrollable factors, which always results in wasted time and money. In the biopharmaceutical industry contamination can cause problems, such as the loss of entire batches. It is thus extremely important to understand how sample contamination can occur and what methods are available to limit and, ultimately, prevent it [41]. Broadly, contaminants are of three types: physical, chemical and biological. Since biological contaminants are a major concern, we will focus on them. Biological contaminants can be divided into two groups:

- Those that are easy to detect in cultures, include bacteria and fungi.
- Those that are more difficult to detect and cause serious problems, include *Mycoplasma*, viruses and cross-contamination by other mammalian cells.

5.5.1.1 Bacterial and fungal contaminants

The ubiquitous nature, small size and fast growth rates of micro-organisms, in particular bacteria, fungi (including molds) and yeasts, makes cultures more vulnerable to attack. These micro-organisms have the tendency to rapidly multiply and propagate in the rich cell culture environment. Bacterial contamination can be detected by microscopic analysis or by their direct effects on the culture (pH shifts, turbidity and cell death). This can only be observed in the absence of antibiotics.

Other micro-organisms such as yeast also affect the growth medium in such a way that it becomes either cloudy or turbid, while fungi develop branched mycelium, which at the end appear as hairy clumps in the medium.

5.5.1.2 Mycoplasma

Mycoplasma (0.15–0.3 μm) are the smallest free-living bacteria-like organisms, and are a genus of bacteria that lacks a cell wall around their cell membrane. They are considered as the most well-known biological contaminant. The P1 antigen is considered to be the main virulence factor of *Mycoplasma.* Due to the lack of a cell wall and small genome, the interaction between the host and *Mycoplasma* is different than for other micro-organisms. *Mycoplasma pneumoniae* is considered to be the most common pathogenic species infecting humans. As *Mycoplasma* lack a cell wall, they are resistant to the action of the host's lysozymes (enzymes that can damage bacterial cell walls). Additionally, due to the lack of a cell wall, most antibiotics active against cell walls are ineffective against *Mycoplasma.* Due to their smaller size they grow rapidly without any visible signs. Thus their small size increases the chance of contamination, as they can sometimes slip through the pores of the filter membranes used in sterilization [42–44].

Mycoplasma are extremely detrimental to any cell culture as contamination causes harmful effects on cellular physiology and metabolism. In contrast to bacteria or fungi, *Mycoplasma* do not cause consistently observable changes in a cell culture, such as rapid pH changes or culture turbidity [45–47]. In addition to causing disturbances in cellular metabolism and morphology, they cause chromosomal aberrations and damage, and can incite cytopathic responses. *Mycoplasma* contamination is more common in European countries than in the US. Even in Japan, the rate of *Mycoplasma* based contamination is high, close to 80%. The discrepancy between the US and the rest of the world is probably due to the use of testing programs. Statistics show that laboratories that routinely test for *Mycoplasma* contamination have a much lower incidence; once detected, contamination can be contained and eliminated. Testing for *Mycoplasma* should be performed at least once per month, and there is a wide range of commercially available kits. The only way to ensure the detection of species is to use at least two different testing methods, such as DAPI staining (DAPI, or 4′,6-diamidino-2-phenylindole, is a fluorescent staining technique that binds intensely to DNA, in particular adenine–thymine rich regions). It is essential to understand the sophisticated mechanisms of *Mycoplasma* that allow survival and virulence under harsh conditions [45–57]. It has been shown that these include mechanisms by which they establish relationships with the host and the virulence is linked to the release of extracellular vesicles [56–64]. In prokaryotes and eukaryotes, the mechanisms of the secretion of extracellular membrane vesicles are involved in establishing a significant part of the bacterial secretome (proteins released into the extracellular space) [65]. In addition to the membrane components, they may contain cytoplasmic proteins, toxins, DNA and RNA [66, 67]. Vesicles play a significant role in building cell–cell communication. These vesicles act as cargo carriers of cell related information [68–74]. By means of proteomic and transcription investigations, it is possible to detect the internalization of these nanostructures. Extracellular vesicles elicit cell reprogramming. These extracellular vesicles stimulate cell–target reprogramming [68–74], which is again controlled by protein traffic. These extracellular vesicles help in understanding:

- Cell–target reprogramming.
- The transfer of virulence determinants.
- Participation in the formation of the host–parasite system.
- The mechanism of resistance against antimicrobials.
- Adaptation to different environmental conditions.

One of the best approaches to detect *Mycoplasma* contamination is by regular examination, in particular by using one or more special procedures, such as:

- Direct growth on broth/agar.
- ELISA, RNA labeling and enzymatic protocols.
- Specific DNA staining and PCR.

However, no single procedure is 100% reliable. Several protocols are used simultaneously to achieve optimal efficiency. Some of the challenges faced by these procedures are discussed in the following.

- Direct culture procedures are comparatively sensitive to most species, but the overall procedure is lengthy (3 weeks).
- Some procedures are costly and less sensitive to non-cultivable species.
- Specific DNA staining methods are comparatively rapid and are easy to perform, but interpretation of their results is difficult, particularly due to decomposition of contaminant nucleic acids.
- The PCR procedure is fast and inexpensive, however, it requires a stringent assay set-up to prevent the chances of false positive or negative results. This can happen due to contamination with target DNA or inhibition of Taq DNA polymerase [75].
- Enzyme based bioconversion of ADP to ATP can be examined by luciferase-containing kits, however, this requires costly luminometric equipment and may produce false negatives owing to decreased enzyme function [74] or to false positives from eukaryotic ATP in the cell culture medium [76].

One of the best approaches to test for *Mycoplasma* is by testing their frequency. As contamination progresses it leads to the development of an environment that displays fast and dominant growth of the contaminant, which can be examined by regularly testing their frequency. One company (InvivoGen) has recently developed a *Mycoplasma* detection kit, PlasmoTest™, which can be directly placed into the incubator facility along with the cells, allowing more frequent testing and therefore helping to decrease the chances of contamination. PlasmoTest™ is a reliable, simple and cost-effective product for the regular examination of *Mycoplasma*. This kit is based on the detection of *Mycoplasma* by engineered cells that express Toll-like receptor 2. These receptors belong to a class of pathogen recognition receptors that detect bacterial membrane components, in this case lipoproteins. Thus, based on this phenomenon, PlasmoTest™ can be utilized to detect *Mycoplasma* in cell cultures.

This universal sensitivity, combined with an easy-to-use format, makes PlasmoTest™ suitable for preventing the risk of contamination [38, 77–81].

5.5.1.3 Viruses

In contrast to contamination with microbes and *Mycoplasma*, viruses never show any signs or cause any changes in the physical characteristics of a culture as they barely affect the pH of the culture medium, which in other cases can result in turbidity. Based on recent reports, it has been realized that certain viruses potentially cause morphological changes to host cells. This phenomenon is called the cytopathic effect and can be detected using microscopic techniques. However, most viral infections result in the introduction of the viral genome as a provirus, which can result in no morphological variations. So ultimately, this type of integration does not affect the phenotypic characteristics of cultured cells infected with a virus. Thus the production of viruses from these cell lines can be a serious threat for other cell cultures and for operators, individuals and patients. The risk of cross contaminations of contaminated culture can also be dangerous for the surrounding environment [82].

Due to the lack of effective procedures for treating infected cell cultures or detecting the presence of viruses in culture, this can cause a serious threat to individuals, in particular the laboratory personnel working on such cultures. As viruses have the tendency to use the host's machinery for replication, the antiviral drugs utilized against them, in particular those with the ability to block viruses, can also be highly toxic for the cells that are being cultured.

5.5.1.4 Other mammalian cell types

Cross-contamination of a cell culture with other cell types is a serious problem that has only recently been considered alarming. An estimated 15%–20% of cell lines currently in use are misidentified, a problem that began with the first human cell line, HeLa, an unusually aggressive cervical adenocarcinoma isolated from Mrs Henrietta Lacks in 1952. HeLa cells are so aggressive that, once accidentally introduced into a culture, they quickly overgrow the original cells. But the problem is not limited to HeLa; there are many examples of cell lines that are characterized as endothelial cells or prostate cancer cells, but are actually bladder cancer cells, and characterized as breast cancer cells but are in fact ovarian cancer cells. In these cases, the problem occurs when the foreign cell type is better adapted to the culture conditions, and thus replaces the original cells in the culture. Such contamination clearly poses a problem for the quality of research produced, and the use of cultures containing the wrong cell types can lead to the retraction of published results. Viral detection was traditionally dependent on the isolation of the virus from cell cultures and this procedure is considered as standard for the laboratory diagnosis of viral disease, however, this method is slow and needs extensive technical expertize. For the fast detection of viral antigens or other viral components, nonculture methods have been developed, such as:

- Centrifugation-enhanced inoculation.
- Co-cultivated cell cultures.

- Cryopreserved cell cultures.
- Newer cell culture formats.
- Precytopathogenic effect detection.
- Transgenic cell lines.

One approach for identifying viral contamination is Fourier transform infrared microspectroscopy, which helps in differentiating viral, bacterial and fungal cell contaminations [83]. Sources of microbial contamination are:

- Airborne particles and aerosols during normal lab work.
- Clothing, which can support and transport a range of micro-organisms.
- Frequently used laboratory equipment, such as microscopes, refrigerators, cold storage rooms and water baths.
- Improperly cleaned and contaminated culture containing incubators.
- Laboratory equipment, such as vortexers, pipetting devices and centrifuges, without biocontainment vessels, can produce large amounts of microbial-laden particulates and aerosols.
- Laboratory personnel (i.e. microbes associated with human bodies).
- Overloading of raw materials in a moist sterilizer (autoclave) can lead to in incomplete sterilization.
- The culture medium itself—improper sterilization of bovine serums, reagents and plasticware.
- Cross-contamination—the chances of cross-contamination are greater when the same person is working simultaneously over multiple cell lines [37, 38, 75, 87–93].

To understand the influence of viral contamination on a cell line, cellular behavior and its physiology should be studied in detail at the molecular level, in particular the membrane potential and viral budding. The membrane potential is generated by the unequal distribution of ions, particularly K^+, Na^+ and Cl^-, across the plasma membrane. This concept can be utilized to measure the membrane potential of a cell under normal conditions. For example, cardiac myocytes typically exhibit a plasma membrane potential of −90 mV [19]. However, this potential will be disturbed by the budding due to viral infection. As the virus or foreign body enters the cell it can disturb the ideal membrane potential in a suitable environment. By using confocal microscopy and fluorophores, such as tetramethylrhodamine methyl ester, the new membrane potential can be determined. The comparison of the resting membrane potential and the change when viral budding starts, can provide a signal which, when measured properly for different classes of cells, provides a way to determine the presence of viral infection (figure 5.5) [86]. The kinetics of virus entry by endocytosis, i.e. the viral budding, can be studied in detail to detect its presence inside a cell [84, 85].

5.6 Aseptic conditions

For animal tissue culture laboratories, it is essential to maintain aseptic conditions, particularly to eliminate all the micro-organisms present in the environment.

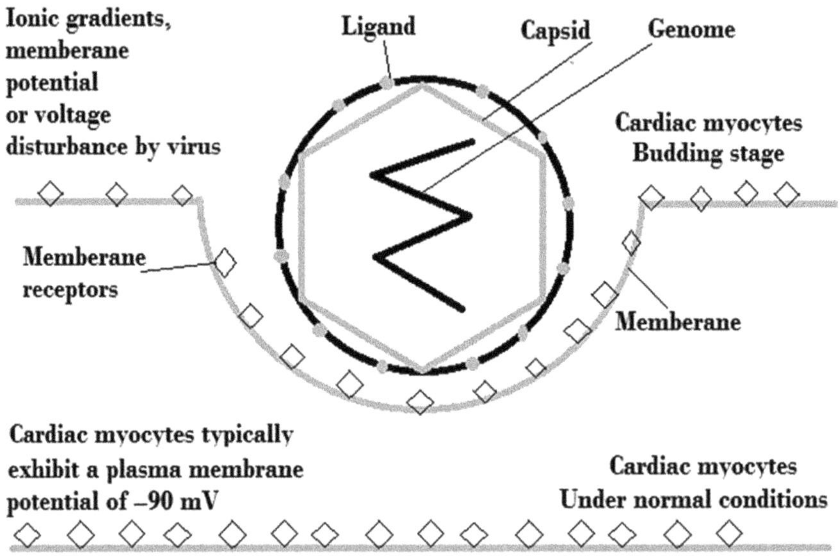

Figure 5.5. Virion entry into the host cell causes voltage disturbance of the membrane.

Different measures have been suggested for the prevention of contamination and maintenance of aseptic conditions, such as:

- Dissection instruments must be sterilized by flame or using dry heat at 180 °C for 2 h.
- The glassware, medium and substrate must be sterilized by moist air sterilization (autoclaving) at a temperature of 121 °C and pressure of 14–15 Psi for 15–20 min. Occasionally, as an alternative to autoclaving, 10% bleach solution can be utilized by soaking glassware or other desired materials and then rinsing with distilled water.
- Hands should be cleaned with disinfectant to the elbows.
- Standard operating procedures for sterile techniques should be followed.
- Membrane filtration of porosity 0.2 μm can be utilized to sterilize heat labile constituents such as serum, trypsin, proteins, growth factors, etc, which should not be autoclaved.
- Regular monitoring of culture for any change in morphological features (by eye or by phase contrast microscope).
- Sterility testing of media and other reagents; this should be always be checked properly before use.
- Laminar air flow hoods need to be sterilized properly before culturing the cells or tissues. Laminar flow hoods and biological safety cabinets are mostly enclosed systems that use filters and directional air flow to provide a contaminant-free work area. The two main types of laminar flow hoods are horizontal and vertical (figure 5.6). Horizontal laminar flow hoods pull air from the environment; the air goes through a filter and then blows smoothly out the front of the hood back into the room. A vertical laminar flow hood works in mostly the same way, but the air is blown from the top of the hood

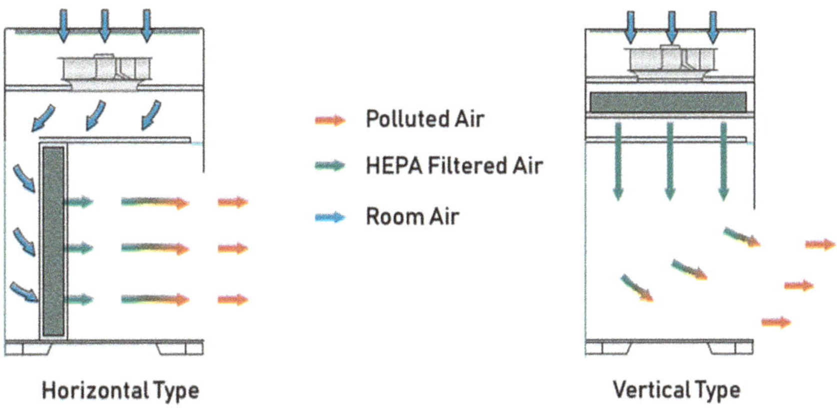

Figure 5.6. Horizontal and vertical laminar air flow system (LAF).

straight down. The working platform of the laminar air flow hood should be initially cleaned with 70% ethanol by first covering the inside area with outward sweeping. Then the glass window or lid should be closed. Subsequently, the ultraviolet lamp must be turned on for 15–20 min. Make sure that body parts, in particular skin and eyes, are not exposed to the UV lamp as it is cancer causing. This UV sterilization must be followed by the main sterilization, performed using high efficiency particulate air (HEPA) filters, which are present in front of the air blower in laminar air flow. The air flow through the HEPA filters should be kept on for at least 5 min to eliminate contaminated air from of the LAF cabinet. Most air purifiers available on the market contain HEPA filters and are based on the same working principal. These ventilator systems, in the form of air purifiers, are used to reduce airborne micro-organisms by generating positive air pressure which replaces contaminated air at the working site [2]. The ventilatory system works by allowing air inflow to be passed through the prefilter (to filter dust particles) followed by the HEPA filters and then to working station. HEPA filters reduce at least 99.97% of airborne particles 0.3 μm in diameter. HEPA filters are the most commonly used ventilatory systems, although others are also available, such as plenum, laminar flow and ex-flow systems (Howorth enclosure).

- The use of the mouth for pipetting liquid medium and cell suspensions should be avoided. As an alternative, bulb type pro-pipettes, pi-pumps, etc, are used to transfer the medium or culture from one vessel to another.

5.7 Sterilization procedures

Sterilization procedures are utilized to kill micro-organisms, however, some sterilization procedures also ensure the *prevention* of micro-organisms, e.g. coating glassware with ethylene oxide. Ethylene oxide based sterilization is the most common industrial sterilization technique used for medical devices. It is a comparatively 'cold' sterilization technique and provides high compatibility with most materials used in the manufacture of medical devices, such as plastics, polymers, metals and glass. Sterilization procedures can broadly be divided into five categories:

- Dry heat sterilization/depyrogenation.
- Moist heat sterilization.
- Filtration.
- Chemical based sterilization.
- Radiation based sterilization.

5.7.1 Dry heat sterilization/depyrogenation

Dry heat sterilization/depyrogenation is usually performed at a temperature of 160 °C for 1 h, however, it involves many procedures for which the temperature range and time duration will vary:

- *Flaming.* The process of sterilizing metallic devices such as needles by exposing them to a flame for few minutes. Flame not only burns the surface micro-organisms, but also burns dust particles on the instrument.
- *Incineration.* Loop inclinators or sterilizers are available to sterilize the inoculating loops used in a microbiology laboratory. The metallic loop present at the distal end of the loop is heated at a high temperature until it is red hot in the flame. This type of sterilization kills all the contaminants present over the surface of the metallic loop.
- *Hot-air oven.* Hot-air oven facilities are used for dry heat sterilization in which suitable materials, such as powders, metal devices, glassware, etc, are exposed to hot air.
- *Radiation sterilization.* The exposure of packed materials to radiation for a certain period of time. This approach involves two types of sterilization: non-ionic and ionic radiation. Non-ionic radiation, including ultraviolet radiation, is safe for the operator and can be used to prevent the entry of live microbes through the air. Ionizing radiation based sterilization uses powerful radiation, such as x-rays, γ-rays, etc, which can be utilized against all types of micro-organisms, however, they have harmful effects on the operator. Thus the operator has to prevent exposure by using specialized clothing. Usually a lead–rubber jacket or plexiglas sheet offer a sufficient amount of protection.

5.7.2 Chemical methods of sterilization

This type of sterilization is usually achieved by converting chemicals into their gaseous state which allows a high penetration of sterilant in the material that needs to be sterilized. As the penetration rate of gases is greater than that of other states, they penetrate rapidly into the material, which makes the method more effective. However, the risk of explosion and cost factors need to be considered. This method involves chemicals that become very poisonous when converted into their gaseous state (figure 5.7). Low temperature based ethylene oxide and formaldehyde treatment are popular. During this procedure, ethylene oxide gas is mixed with carbon dioxide to minimize the chances of an explosion. Similarly, hydrogen peroxide plasma based sterilization is also utilized for the same purpose.

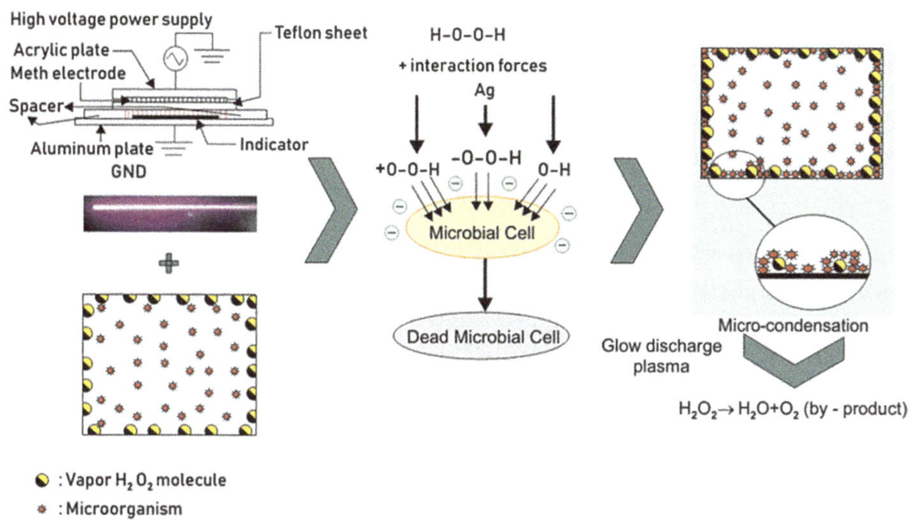

Figure 5.7. Chemical sterilization.

5.7.3 Filtration

This method is utilized to sterilize or filter contaminants, in particular microbes, from liquids that are heat-sensitive in nature by using different types of filters. The risk of clogging and the long duration of the process are the considerable disadvantages of this procedure. Different filter papers or membranes of pore size 0.2–0.22 μm (e.g. hydrophobic PTFE membrane filters made of polypropylene with a pore size of 0.2 μm) are used, however, this range is not enough to filter contaminants smaller than 0.2 μm, thus 0.1 μm membranes have been developed [94]. However, these minimum size membrane (0.1 μm or smaller) filters result in lower flow rates and reduced throughputs [94]. Alternatively, a 0.45 μm filter can be utilized to remove fungal spores. An ideal size range for filters used exclusively for sterilization is 0.1–0.2 μm. Many filters, made of nylon, cellulose nitrate, polycarbonate and ceramics, are available on the market. These filters are available in various shapes and sizes with different designs, e.g. disc filters, cartridges, syringe filters and hollow fibers [94]. Three types of filters are discussed in more detail in the following sections.

5.7.3.1 Membrane filters

Membrane filters are thin membrane filters which are made of cellulose and are highly efficient in sterilizing liquids, solvents and gasses. They can be used for sterilization purposes during injection by introducing a membrane between the syringe and needle. One of the major drawbacks of membrane filters is the risk of disruption to the membrane resulting in improper sterilization.

5.7.3.2 Seitz filters

Seitz filters are made of layers of asbestos or other materials, forming a pad-like structure, thicker than membrane filters. One of the main advantages of these filters is that they do not rupture during filtration, however, the chances of absorption of solution by the pad surface are greater. Another type of filter, which can be used instead of Seitz filters, is the sintered glass filter, which is fabricated from glass and thus does not absorb liquids during filtration. However, the main disadvantage of these materials is that they are very brittle and break easily.

5.7.3.3 Candle filters

Candle filters are ceramic, fabricated using special clay-like diatomaceous mud (made of algae shells) which offers minute pores that can potentially filter many tiny particles and micro-organisms. During this type of filtration, micro-organisms are trapped in the matrix due to the densely packed arrangement of the candle.

5.7.4 Moist heat sterilization

Certain fluids and perishable materials can be sterilized using moist heat sterilization at 121 °C for 15–20 min. For effective sterilization it is essential that the steam penetrates all parts of the sterilizing materials. Some heat labile chemicals cannot be sterilized, e.g. urea cannot be sterilized by autoclaving in combination with other chemicals as it can decompose during heating, which can result in the formation of CO_2 and NH_3. To avoid this, urea can be filter sterilized and added aseptically to autoclaved media. Similarly, autoclaving is not suitable for sterilizing sulfur as it melts the sulfur powder, so that it is no longer available for micro-organisms. Modern autoclaves can be utilized to sterilize different types of oils. The methods used for sterilization of equipment, apparatus and liquids are listed in table 5.2.

Table 5.2. Methods of sterilization for equipment, apparatus and liquids.

Materials	Methods
Surgical instruments	Boiling, incineration,] and autoclave
Glassware	Autoclave, boiling and hot-air oven
Water and liquids	Filtration; for liquids other than water also the autoclave
Powders and other dry forms	Hot-air oven if thermostable, otherwise gaseous methods and radiation
Chemicals	Filtration
Culture media	Autoclaving, but it should not contain any heat labile substances
Heat-sensitive devices	Hydrogen peroxide sterilization or plasma sterilization, kills bacteria, viruses, fungi and even bacterial spores
	Systems using low temperature hydrogen peroxide gas plasma technology (STERRAD®) to sterilize a wide range of instruments.
Air sterilization	HEPA filters with LAF, radiation based sterilization, ozone based sterilization and low temperature based sterilization (ethylene oxide and formaldehyde based sterilization).

Recently, ultrasound waves have been tested for their sterilization effect against different contaminants by checking the shelf-life and quality parameters of ultrasound treated fruit and vegetable juices, and it was observed that the ultrasound treated juice satisfied the microbiological and physiochemical safety parameters in refrigerated storage conditions for 20 days, using methods suitable for large-scale processing.

Ultrasound waves are not as effective as other procedures, however, their utilization is found to be useful in tissue cultures. All these methods require an ideal approach to check the effectiveness of the sterilization procedure. To check whether 100% sterilization has been achieved or not, or to check the effectiveness of the procedure used for the sterilization process, specific types of bacterial spores have been suggested [95]. These spores are biological indicators for in-process validation or authentication of the process. It is well known that spores are used because most spores are highly resistant against heat, radiation and sterilant, thus they can be used as different types of biological indicators to check diverse sterilization procedures. Numerous biological indicators offer an outstanding platform to authenticate or validate different sterilization processes. The performance of a biological indicator is influenced by both the inherent features of the organism as well as environmental factors before and after their production. Endospores have a wide variety of important medical and industrial applications. The life cycle, spore structure and factors that influence the spore resistance of spore-forming bacteria should be studied in detail (figure 5.8) [95, 96].

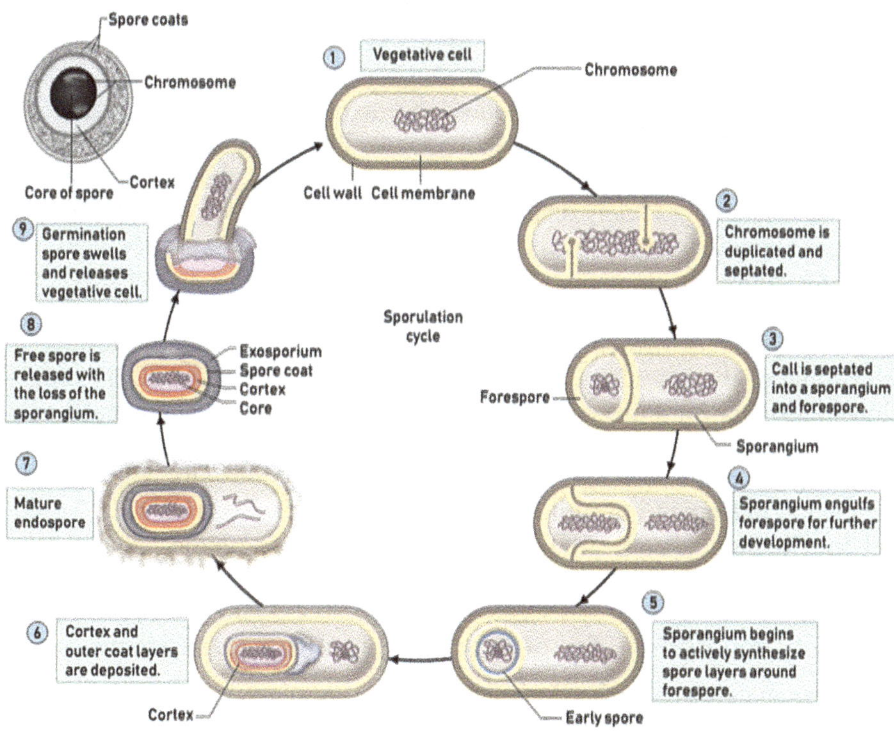

Figure 5.8. Life cycle and spore structure of spore-forming bacteria.

5.8 Advantages and disadvantages of animal tissue culture

It is now clear that cell culture technology has certain advantages that offer excellent solutions to the challenges faced during the assessment of *in vivo* assays using certain biological models. It is essential to discuss the recent advancements and prospects of cell culture technology, and also the remaining disadvantages. The effects and metabolism of various toxic compounds and drugs can be assessed using cell culture technology. This type of assessment can be performed using cells present in physiologically complex tissue (by culturing in optimal conditions) to test the effect of drugs and toxic compounds. Several factors, such as the growth rate, the ingredients of the culture medium, the culture conditions and the population density, can be regulated. In addition, the genetic make-up of cells can be manipulated using transfection to study the function of various genes. The effects of toxic molecules, carcinogenic agents, drug interactions and biopharmaceuticals can be investigated by cell culture technology. Also, clones can be developed and studied from heterogeneous cell populations. Cell or tissue function can also be tested using animal tissue culture science. Various assays, such as immunoelectron, molecular, biological, microscopical and immunohistochemical assays, can also be used to obtain more information about new medicines before their use in humans [97]. Cell differentiation and other physiological aspects, in particular molecular mechanisms, can also be explored with this technology. Using current animal tissue culture science, desirable cells can be isolated, reproduced and manipulated to derive certain cells which can be utilized for multiple purposes, such as in transplantation or regenerative science for use against certain diseases. To meet the performance, quality, precision and safety standards the ingredients of cell based assays should be selected carefully. It was believed previously that animal cells could be cultured under *in vitro* conditions by using a Petri dish or flask, just like bacteria and other micro-organisms, provided that optimal conditionals for growth were provided. This earlier belief gradually changed when it materialized that these culture technologies considerably changed the cellular make-up of cultured cells. Based on current reports, it has been suggested that the cellular environment should be considered as 'instructions' for the development of the cell. Mimicking the instructions of the *in vivo* environment in *in vitro* conditions always results in the loss of some of the key instructions, which can lead to certain cellular injuries or malfunctions later on. One single contaminant can affect these systematic instructions, which can eventually ruin the whole culture. Thus optimized procedures and high-quality reagents are required to improve the uniformity and validity of experimental outcomes and ultimately to produce quality-assured cells. To understand this in more detail, we should first be familiar with the advantages and disadvantages of animal tissue culture, as listed in table 5.3.

5.9 Safety regulations for animal tissue culture laboratories

The regular maintenance of cell cultures and other experimental work can expose researchers to risks. These can be related to the equipment, reagents, chemicals, sterilization procedures and the tissue itself. In addition, there are risks associated

Table 5.3. Advantages and disadvantages of animal tissue culture.

Advantages	Disadvantages
Controlled environment (physico-chemical parameters such as temperature, pH, dissolved gas and osmolality).	Maintaining controlled environmental conditions during animal tissue culture experiments is not easy.
Physico-chemical parameters can be controlled by nutrient concentration, cell–cell interactions and hormonal control.	Most animal tissue culture facilities, including infrastructure, equipment, chemicals and apparatus, are expensive, which ultimately increases the cost of the entire process. The cost of production through animal tissue culture is ten times greater than the direct use of animal tissue.
Homogeneous cells can be developed after 2–3 successive cultures. These homogeneous cell-containing culture can be utilized for a variety of purposes.	The continuous cell line results in the genetic instability of cells, which can eventually results in the heterogeneity of the cells.
Characterization of the cells, in particular cytological and immunological characterization, can be achieved.	Professional and skilled people are required for the development and use of cultures.
Cell lines can be maintained under liquid nitrogen for many years.	The components for maintaining homeostatic conditions, in particular for maintaining cellular programming, are lacking under *in vitro* conditions. Moreover, it is challenging to understand the molecular role of growth factors and hormones in this process.
Many different studies, such as drug toxicity, cell–cell communication, cellular programming, the development of cells under different environments, diseased cells (in particular cancer cells) and cellular mechanics, can be carried out, and the effects of different treatments such as peptides, proteins, nucleic acids and other biomolecules on cells can be investigated.	Due to differences in *in vitro* and *in vivo* environments, cells always lose their originality. This mainly occurs when the first culture is established.
Owing to direct access to the medium, the cells grow quickly, in particular when using certain feeder layer techniques or growth promoters. Smaller amounts of chemicals are required in comparison to *in vivo* models, where most chemicals are	2D and 3D animal tissue culture environments are still unable to 100% mimic the *in vivo* cellular environment. Since under *in vivo* conditions all cells exist in a 3D environment, it is essential to develop culture models or technique to mimic *in vivo* conditions.

(*Continued*)

lost or wasted through distribution to various tissue and excretion.	
Animal utilization can be drastically reduced or prevented.	Most primary cell cultures have a limited life span (a limited number of passages).
Cell lines grow comparatively faster and need minimal care.	Cell lines may behave differently in contrast to primary cells (the first culture) in response to stress.
Using molecular techniques primary cells can be transformed and can be subcultured for unlimited passages.	Tracking genetic changes, mutations and changes in cellular programing and metabolism is challenging.
With the aid of cellular imaging and fluorescent tagging methods, many physiological and molecular events can be studied.	High chances of genetic variation.
Most bioconversion and biosynthetic pathways can be elucidated.	Maintaining aseptic conditions is challenging.
Can be utilized to produce vaccines and monoclonal antibodies	The microenvironment in the culture vessel can cause physical, chemical and physiological changes.
Cytotoxicity studies can be performed to check the toxicity of different drugs.	Require standardization of the medium, concentration of nutrients and serum.

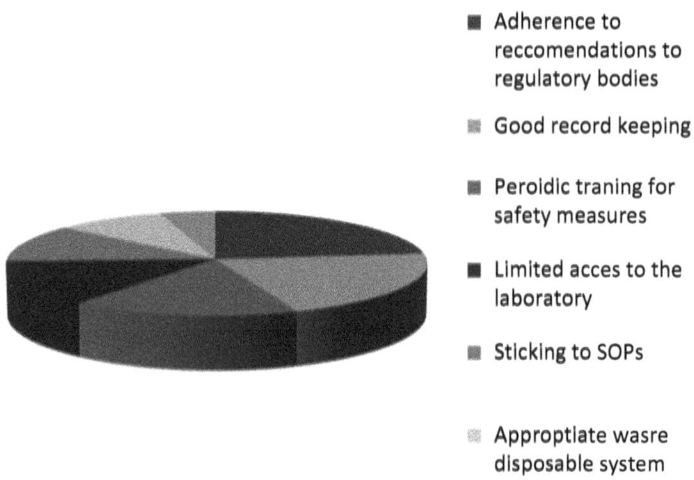

Figure 5.9. Safety regulations for animal tissue culture laboratories.

with the procedures utilized to alter normal cell function, many of which can have harmful effects. Risks can be of different origins, such as those associated with chemicals, radiations, biological tissues, cells and mechanical components. Despite their familiarity with the involved risks, researchers need to take care to work

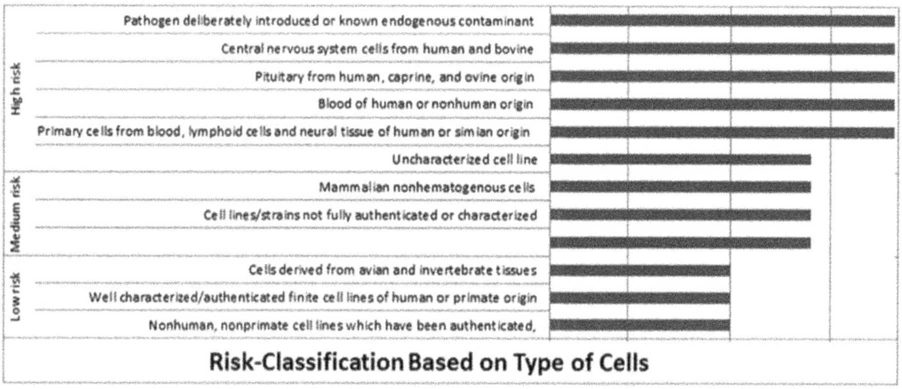

Figure 5.10. Risks associated with types of cells.

efficiently and keep safe. Proper training should be provided to the researchers; familiarity with the risks associated with animal tissue culture environments and safety guidelines is essential. By using these guidelines, the risk to researchers can be reduced. This brief section covers some safety considerations in the tissue culture laboratory.

Animal tissue culture research laboratories have several specific hazards, mainly associated with handling and manipulating human or animal cells and tissues, as well as toxic, corrosive and mutagenic solvents and reagents. Most accidents occur due to ignorance and casual approaches to working with radiological and biological samples. Risk evaluation is an efficient method to categorize risks, assess hazards, and introduce suitable procedures to alleviate hazards for any work procedure (figure 5.10). During cell based *in vitro* procedures, the hazard evaluation includes the evaluation of potential exposure to risk arising from the handling of harmful reagents, chemicals, solvents cell cultures and micro-organisms, and to stop lab-originated infections of different micro-organisms (bacteria, viruses, fungi, *Mycoplasma*), primary or secondary (unfixed) tissues, human cell lines, human blood and rDNA. Several devices and approaches, such as laminar air flow systems, air humidifiers, biosafety cabinets, sticking to good laboratory practices, management of the use of various cell lines and better administrative controls (organizational policies), in particular sticking to national and global guidelines for risk assessment (figure 5.9) make up a safe animal cell culture laboratory. For risk assessment, the following essential measures should be considered [98]:

- Timely reviews in order to avoid future occurrences of any risks.
- Thorough communication of risk assessments to staff and careful maintenance of facilities.
- A rigorous process of documentation.
- Detailed prior planning.

One of the most serious biohazards presented by cell cultures are those viruses that are pathogenic to humans. Viral contaminants that cause infection in primate tissues can also cause human disease, and are thus considered as potential

Table 5.4. Viral contamination.

Viral contamination	
Main viral contaminants in human tissue	
Hepatitis viruses	HBV (hepatitis B virus)
	HCV (hepatitis C virus)
	HDV (hepatitis D virus)
	HEV (hepatitis E virus)
Human retroviruses	HIV-1
	HIV-2
	HTLV-1
	HTLV-2
	SIV
Herpes viruses	EBV
	CMV (cytomegalus virus)
	Human herpes virus-6
	Human simian virus 1 and 2
Popoviruses	Different human papillomaviruses (HPV)
Prions	Infectious prion proteins
Viral contaminants in primate tissues causing human disease	
RNA viruses	Flaviviruses
	Filoviruses
	Simian hemorrhagic virus
	Rabies virus
	HAV
	Poliovirus
DNA viruses	Herpes viruses (herpes-b virus and others)
	Mossuscum contagiosum
	Lymphocytic choriomeningitis virus
Mouse viruses pathogenic for humans	
	Hantan virus (hemorrhagic fever with renal syndrome)
	Sendai virus (murine parainfluenza virus type 1)

contaminants (table 5.4). Cell cultures which are contaminated with viruses signify a considerable danger for operators, collaborators and patients, as well as non-contaminated cell cultures. Most viral strains that cause viral infections are highly pathogenic, and suitable treatment is not available for many viruses, thus the prevention of viral infections should be considered (table 5.4). In comparison to other infections, in particular bacterial and fungal infections, viral contamination cannot be identified easily as viruses are too small to be detected by normal light microscopy (table 5.4). The main indicators of viral contamination in cell culture are: (a) changes in morphology and (b) full genomic sequencing. It is important to remember that cell lines are not 'virus-free'—they were isolated from living samples and so it is a given that all cell lines have viruses in them—typically these viruses are adapted to grow in their environment without showing cytopathic effects and so do not cause a problem. The real problems occur when viruses are introduced due to poor aseptic techniques. As a rule of thumb, as long as you do not see any changes in morphology or culture dynamics, any contaminant can be considered as incidental. If you have a contaminant, you have to go back to low passage stock or buy in new stock from an accredited culture collection such as ATCC or ECACC. As mentioned above, if viral contamination results in morphological changes in the cultured cells, such as cytopathic effects, then contamination by a virus can be suspected. In most cases viral contamination is silent in nature, with no noticeable morphological modifications in the virus-infected culture. This type of infection is of great concern. Occasionally, virus-infected cells show changes in their vulnerability against infection by other viruses, e.g. certain safety testing procedures use indicator cells to detect if cells are chronically infected by viruses, which decrease their vulnerability against other virus species. However, this can result in false negative results if the virus to be identified can no longer infect the indicator cells [99]. Generally cell lines that are infected by viruses cannot be treated to become virus-free. Eventually, potentially valuable cell lines have to be discarded and replaced by new cells. However, there are some exceptions to this rule, such as the presence of lactate dehydrogenase. This virus cannot infect the cells and it will be diluted during subsequent *in vitro* passaging, and thus will be lost [100, 101]. Most developed nations have established safety guidelines to reduce the risks associated with tissue culture laboratories [102, 103] such as 'Biosafety in microbiology and biomedical laboratories' (US Department of Health and Human Sciences, 1993) and 'Safe

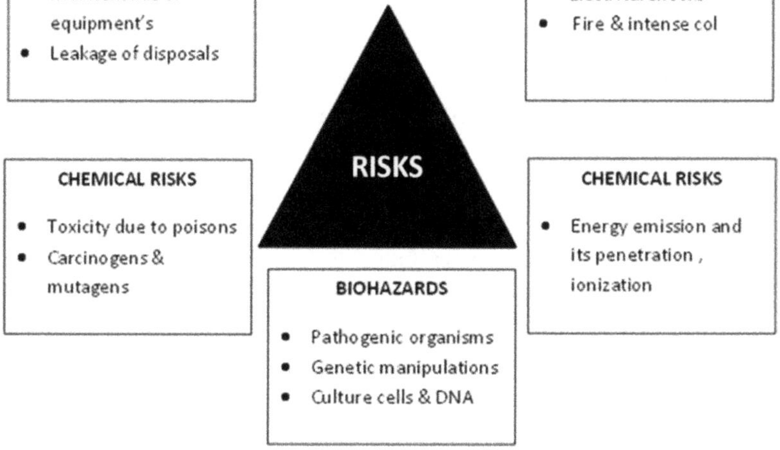

Figure 5.11. Risks and contributory factors.

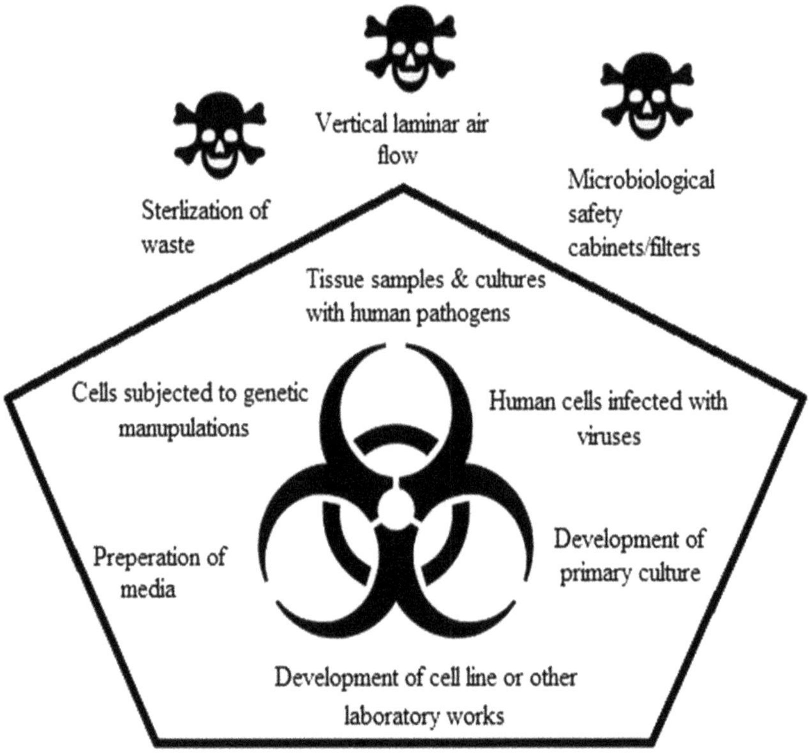

Figure 5.12. The inner area shows sources that contribute to biohazards and the outer area shows suitable control measures.

working and prevention of infection in clinical laboratories' (UK Health Services Advisory Committee, 1991).

Most of the hazards in animal tissue culture practices are due to negligence. The risks and the contributory factors are broadly categorized in figure 5.11.

5.9.1 Biohazards

A risk associated with biological materials is called a biohazard. Most of the risks associated with biohazards are shown in figure 5.12.

References

[1] Kulkarni S, Becker L and Pasricha P J 2012 Stem cell transplantation in neurodegenerative disorders of the gastrointestinal tract: future or fiction? *Gut* **61** 613–21

[2] Chen A K, Reuveny S and Oh S K 2013 Application of human mesenchymal and pluripotent stem cell microcarrier cultures in cellular therapy: achievements and future direction *Biotechnol. Adv.* **31** 1032–46

[3] Kirouac D C and Zandstra P W 2008 The systematic production of cells for cell therapies *Cell Stem Cell* **3** 369–81

[4] Rowley J *et al* 2012 Meeting lot-size challenges of manufacturing adherent cells for therapy *BioProcess Int.* **10** 16–22

[5] van Wezel A L 1967 Growth of cell-strains and primary cells on micro-carriers in homogeneous culture *Nature* **216** 64–5
[6] Clark J, Hirstenstein H and Gebb C 1980 Critical parameters in the microcarrier culture of animal cells *Dev. Biol. Stand.* **46** 117–24
[7] Hirtenstein M, Clark J, Lindgren G and Vretblad P 1980 Microcarriers for animal cell culture: a brief review of theory and practice *Dev. Biol. Stand.* **46** 109–16
[8] van Wezel A L 1971 New trends in the preparation of cell substrates for the production of virus vaccines *Prog. Immunobiol. Stand.* **5** 187–92
[9] van der Velden-de Groot C A 1995 Microcarrier technology, present status and perspective *Cytotechnology* **18** 51–6
[10] Nilsson K, Buzsaky F and Mosbach K 1986 Growth of anchorage-dependent cells on macroporous microcarriers *Nat. Biotechnol.* **4** 989–90
[11] Varani J *et al* 1985 Substrate-dependent differences in growth and biological properties of fibroblasts and epithelial cells grown in microcarrier culture *J. Biol. Stand.* **13** 67–76
[12] Reuveny S *et al* 1982 DE-52 and DE-53 cellulose microcarriers. I. Growth of primary and established anchorage-dependent cells *In Vitro* **18** 92–8
[13] van Wezel A L and van Steenis G 1978 Production of an inactivated rabies vaccine in primary dog kidney cells *Dev. Biol. Stand.* **40** 69–75
[14] Seefried von A and Chun J H 1981 Serially sub cultivated cells as substrates for poliovirus production for vaccine *Dev. Biol. Stand.* **47** 25–33
[15] Meignier B 1979 Cell culture on beads used for the industrial production of foot-and-mouth disease virus *Dev. Biol. Stand.* **42** 141–45
[16] Giard D J *et al* 1979 Human interferon production with diploid fibroblast cells grown on microcarriers *Biotechnol. Bioeng.* **21** 433–42
[17] Mignot G *et al* 1990 Production of recombinant Von Willebrand factor by CHO cells cultured in macroporous microcarriers *Cytotechnology* **4** 163–71
[18] Costa A R, Withers J, Rodrigues M E, McLoughlin N, Henriques M, Oliveira R, Rudd P M and Azeredo J 2013 The impact of microcarrier culture optimization on the glycosylation profile of a monoclonal antibody *Springerplus* **2** 25
[19] Ojala M, Rajala K, Pekkanen-Mattila M, Miettinen M, Huhtala H and Aalto-Setälä K 2012 Culture conditions affect cardiac differentiation potential of human pluripotent stem cells *PLoS One* **7** e48659
[20] Griffon D J, Sedighi M R, Sendemir-Urkmez A, Stewart A A and Jamison R 2005 Evaluation of vacuum and dynamic cell seeding of polyglycolic acid and chitosan scaffolds for cartilage engineering *Am. J. Vet. Res.* **66** 599–605
[21] Wang L, Tran I, Seshareddy K, Weiss M L and Detamore M S 2009 A comparison of human bone marrow-derived mesenchymal stem cells and human umbilical cord-derived mesenchymal stromal cells for cartilage tissue engineering *Tissue Eng. Part* A **15** 2259–66
[22] Mao X, Chu C L, Mao Z and Wang J J 2005 The development and identification of constructing tissue engineered bone by seeding osteoblasts from differentiated rat marrow stromal stem cells onto three-dimensional porous nano-hydroxylapatite bone matrix *in vitro* *Tissue Cell* **37** 349–57
[23] Pedraza C E, Marelli B, Chicatun F, McKee M D and Nazhat S N 2010 An *in vitro* assessment of a cell-containing collagenous extracellular matrix-like scaffold for bone tissue engineering *Tissue Eng. Part* A **16** 781–93

[24] Kruyt M C, Dhert W J, Yuan H, Wilson C E, van Blitterswijk C A, Verbout A J and de Bruijn J D 2004 Bone tissue engineering in a critical size defect compared to ectopic implantations in the goat *J. Orthop. Res.* **22** 544–51

[25] Kim K, Dean D, Lu A, Mikos A G and Fisher J P 2011 Early osteogenic signal expression of rat bone marrow stromal cells is influenced by both hydroxyapatite nanoparticle content and initial cell seeding density in biodegradable nanocomposite scaffolds *Acta Biomater.* **7** 1249–64

[26] Xu H H, Quinn J B, Takagi S and Chow L C 2004 Synergistic reinforcement of *in situ* hardening calcium phosphate composite scaffold for bone tissue engineering *Biomaterials* **25** 1029–37

[27] Cooke M J, Phillips S R, Shah D S H, Athey D, Lakey J H and Przyborski S A 2008 Enhanced cell attachment using a novel cell culture surface presenting functional domains from extracellular matrix proteins *Cytotechnology* **56** 71–9

[28] Kumar A, Nune K C, Basu B and Misra R D 2016 Mechanistic contribution of electroconductive hydroxyapatite–titanium disilicide composite on the alignment and proliferation of cells *J. Biomater. Appl.* **30** 1505–16

[29] Jeong S I, Jun I D, Choi M J, Nho Y C, Lee Y M and Shin H 2008 Development of electroactive and elastic nanofibers that contain polyaniline and poly(L-lactide-co-epsilon-caprolactone) for the control of cell adhesion *Macromol. Biosci.* **8** 627–37

[30] Tang Y, Wu C, Wu Z, Hu L, Zhang W and Zhao K 2017 Fabrication and *in vitro* biological properties of piezoelectric bioceramics for bone regeneration *Sci. Rep.* **7** 43360

[31] Choi W J *et al* 2015 Effects of substrate conductivity on cell morphogenesis and proliferation using tailored, atomic layer deposition-grown ZnO thin films *Sci. Rep.* **5** 9974

[32] Özgür Ü *et al* 2005 A comprehensive review of ZnO materials and devices *J. Appl. Phys.* **98** 1301

[33] Choi W J *et al* 2015 Effects of substrate conductivity on cell morphogenesis and proliferation using tailored, atomic layer deposition-grown ZnO thin films *Sci. Rep.* **5** 9974

[34] Sanders E R 2012 Aseptic laboratory techniques: plating methods *J. Vis. Exp.* **63** e3064

[35] Coté R J 2001 Sterilization and filtration *Curr. Protoc. Cell Biol.* **1** 1.4.1–1.4.21

[36] Rajpal D K, Qu X A, Freudenberg J M and Kumar V D 2014 Mining emerging biomedical literature for understanding disease associations in drug discovery *Methods Mol. Biol.* **1159** 171–206

[37] Carlson B 2010 Cell culture products ride industry momentum *Genet. Eng. Biotechnol. News* **30**(2) 15 January

[38] Lincoln C K and Gabridge M G 1998 Cell culture contamination: sources, consequences, prevention, and elimination *Methods Cell. Biol.* **57** 49

[39] Stacey G N 2011 Cell culture contamination *Methods Mol. Biol.* **731** 79–91

[40] Uphoff C C and Drexler H G 2011 Elimination of mycoplasmas from infected cell lines using antibiotics *Methods Mol. Biol.* **731** 105–14

[41] Volokhov D V, Graham L J, Brorson K A and Chizhikov V E 2011 *Mycoplasma* testing of cell substrates and biologics: review of alternative non-microbiological techniques *Mol. Cell. Probes* **25** 69–77

[42] Bredt W 1976 Pathogenicity factors of *Mycoplasmas Infection* **4** 9–12

[43] Savage D C, Dubos R and Schaedler R W 1968 The gastrointestinal epithelium and its autochthonous bacterial flora *J. Exp. Med.* **127** 67–76

[44] Chernova O A, Medvedeva E S, Mouzykantov A A, Baranova N B and Chernov V M 2016 Mycoplasmas and their antibiotic resistance: the problems and prospects in controlling infections *Acta Nat.* **8** 24–34
[45] Uphoff C C and Drexler H G 2005 Eradication of *Mycoplasma* contaminations *Methods Mol. Biol.* **290** 25–34
[46] Kenny G E and Pollock M E 1963 Mammalian cell cultures contaminated with pleuro-pneumonia-like organisms. I. Effect of pleuropneumonia-like organisms on growth of established cell strains *J. Infect. Dis.* **112** 7–16
[47] Hearn H J Jr, Officer J E, Elsner V and Brown A 1959 Detection, elimination, and prevention of contamination of cell cultures with pleuropneumonia-like organisms *J. Bacteriol.* **78** 575–82
[48] Catrein I and Herrmann R 2011 The proteome of *Mycoplasma pneumoniae*, a supposedly 'simple' cell *Proteomics* **11** 3614–32
[49] Yus E *et al* 2009 Impact of genome reduction on bacterial metabolism and its regulation *Science* **326** 1263–8
[50] Sugio A, MacLean A M, Kingdom H N, Grieve V M, Manimekalai R and Hogenhout S A 2011 Diverse targets of phytoplasma effectors: from plant development to defense against insects *Annu. Rev. Phytopathol.* **49** 175–95
[51] Miura C, Komatsu K, Maejima K, Nijo T, Kitazawa Y, Tomomitsu T, Yusa A, Himeno M, Oshima K and Namba S 2015 Functional characterization of the principal sigma factor RpoD of phytoplasmas via an *in vitro* transcription assay *Sci. Rep.* **5** 11893
[52] Zhang K, Zheng S, Yang J S, Chen Y and Cheng Z 2013 Comprehensive profiling of protein lysine acetylation in *Escherichia coli J. Proteome Res.* **12** 844–51
[53] Himmelreich R, Plagens H, Hilbert H, Reiner B and Herrmann R 1997 Comparative analysis of the genomes of the bacteria *Mycoplasma pneumoniae* and *Mycoplasma genitalium Nucl. Acids Res.* **25** 701–12
[54] Hopfe M, Deenen R, Degrandi D, Köhrer K and Henrich B 2013 Host cell responses to persistent mycoplasmas—different stages in infection of HeLa cells with *Mycoplasma hominis PLoS One* **8** e54219
[55] Chernov V M, Chernova O A, Medvedeva E S, Mouzykantov A A, Ponomareva A A, Shaymardanova G F, Gorshkov O V and Trushin M V 2011 Unadapted and adapted to starvation *Acholeplasma laidlawii* cells induce different responses of *Oryza sativa*, as determined by proteome analysis *J. Proteomics* **74** 2920–36
[56] Párraga-Niño N, Colomé-Calls N, Canals F, Querol E and Ferrer-Navarro M 2012 A comprehensive proteome of *Mycoplasma genitalium J. Proteome Res.* **11** 3305–16
[57] Liu W *et al* 2012 Comparative genomics of *Mycoplasma*: analysis of conserved essential genes and diversity of the pan-genome *PLoS One* **7** e35698
[58] Chernov V M, Chernova O A, Medvedeva E S, Mouzykantov A A, Ponomareva A A, Shaymardanova G F, Gorshkov O V and Trushin M V 2011 Unadapted and adapted to starvation *Acholeplasma laidlawii* cells induce different responses of *Oryza sativa*, as determined by proteome analysis *J. Proteomics* **74** 2920–36
[59] Razin S and Hayflick L 2010 Highlights of *Mycoplasma* research—an historical perspective *Biologicals* **38** 183–90
[60] Chernov V M, Chernova O A, Gorshkov O V, Baranova N B, Mouzykantov A A, Nesterova T N and Ponomareva A A 2013 Interaction between mycoplasmas and plants:

extracellular membrane vesicles and phytopathogenicity of *Acholeplasma laidlawii* PG8 *Biochem. Biophys.* **450** 483–7

[61] Chernov V M *et al* 2012 Extracellular membrane vesicles and phytopathogenicity of *Acholeplasma laidlawii* PG8 *Sci. World J.* **2012** 315474

[62] Lazarev V N *et al* 2011 Complete genome and proteome of *Acholeplasma laidlawii J. Bacteriol.* **193** 4943–53

[63] Chernov V M, Chernova O A, Medvedeva E S and Davydova M N 2011 Adaptation of mycoplasmas to environmental conditions: features of the proteome shift in *Acholeplasma laidlawii* PG8 at persistent exposure to stressors *Dokl. Biochem. Biophys* **438** 134–7

[64] Chernov V M *et al* 2011 Extracellular vesicles derived from *Acholeplasma laidlawii* PG8 *Sci. World J* **111** 120–30

[65] Kulp A and Kuehn M J 2010 Biological functions and biogenesis of secreted bacterial outer membrane vesicles *Annu. Rev. Microbiol.* **64** 163–84

[66] Lee E Y, Choi D S, Kim K P and Gho Y S 2008 Proteomics in gram-negative bacterial outer membrane vesicles *Mass Spectrom. Rev.* **27** 535–55

[67] Deatherage B L and Cookson B T 2012 Membrane vesicle release in bacteria, eukaryotes, and archaea: a conserved yet underappreciated aspect of microbial life *Infect. Immun.* **80** 1948–57

[68] Gaudin M, Gauliard E, Schouten S, Houel-Renault L, Lenormand P, Marguet E and Forterre P 2013 Hyperthermophilic archaea produce membrane vesicles that can transfer DNA *Environ. Microbiol. Rep.* **5** 109–16

[69] Lee E Y *et al* 2009 Gram-positive bacteria produce membrane vesicles: proteomics-based characterization of *Staphylococcus aureus*-derived membrane vesicles *Proteomics* **9** 5425–36

[70] Amano A, Takeuchi H and Furuta N 2010 Outer membrane vesicles function as offensive weapons in host-parasite interactions *Microb. Infect.* **12** 791–8

[71] Yoon H, Ansong C, Adkins J N and Heffron F 2011 Discovery of *Salmonella* virulence factors translocated via outer membrane vesicles to murine macrophages *Infect. Immun.* **79** 2182–92

[72] Schertzer J W and Whiteley M 2013 Bacterial outer membrane vesicles in trafficking, communication and the host-pathogen interaction *J. Mol. Microbiol. Biotechnol.* **23** 118–30

[73] Kluge S, Hoffmann M, Benndorf D, Rapp E and Reichl U 2012 Proteomic tracking and analysis of a bacterial mixed culture *Proteomics* **12** 1893–901

[74] Pollak C N, Delpino M V, Fossati C A and Baldi P C 2012 Outer membrane vesicles from *Brucella abortus* promote bacterial internalization by human monocytes and modulate their innate immune response *PLoS One* **7** e50214

[75] Drexler H G and Uphoff C C 2002 *Mycoplasma* contamination of cell cultures: incidence, sources, effects, detection, elimination, prevention *Cytotechnology* **39** 75–90

[76] Razin S 2006 The genus *Mycoplasma* and related genera (class Mollicutes) *Prokaryotes* ed M Dworkin *et al* 4 (Berlin: Springer) pp 836–904

[77] McGarrity G *et al* 1992 Mycoplasmas and tissue culture cells *Mycoplasmas, Molecular Biology and Pathogenesis* ed J Maniloff, R N McElhaney, L R Finch and J B Baseman (Washington, DC: American Society for Microbiology) pp 445–54

[78] Razin S *et al* 1998 Molecular biology and pathogenicity of *Mycoplasma Microbiol. Mol. Biol. Rev.* **62** 1094–156

[79] Doyle A and Griffiths J B 1998 The cell: selection and standardization *Cell and Tissue Culture: Laboratory Procedures in Biotechnology* ed A Doyle and J B Griffiths (New York: Wiley) pp 35–52
[80] Uphoff C C *et al* 2014 Detection of *Mycoplasma* contamination in cell cultures *Curr. Protoc. Mol. Biol.* **106** 4
[81] Molla Kazemiha V *et al* 2012 Sensitivity of biochemical test in comparison with other methods for the detection of *Mycoplasma* contamination in human and animal cell lines stored in the National Cell Bank of Iran *Cytotechnology* **66** 861–73
[82] Merten O W 2002 Virus contaminations of cell cultures—a biotechnological view *Cytotechnology* **39** 91–116
[83] Leland D S and Ginocchio C C 2007 Role of cell culture for virus detection in the age of technology *Clin. Microbiol. Rev.* **20** 49–78
[84] Marsh M and Helenius A 2006 Virus entry: open sesame *Cell.* **124** 729–40
[85] Lakadamyali M, Rust M J and Zhuang X 2004 Endocytosis of influenza viruses *Microbes Infect.* **6** 929–36
[86] Zhdanov V P 2015 Kinetics of virus entry by endocytosis *Phys. Rev.* E **91** 042715
[87] Agarwal P and Searls D B 2009 Can literature analysis identify innovation drivers in drug discovery? *Nat. Rev. Drug Discov.* **8** 865
[88] Ryan J 2008 *A Guide to Understanding and Managing Cell Culture Contamination* Corning Life Sciences Technical Literature
[89] Armstrong S E, Mariano J A and Lundin D J 2010 The scope of *Mycoplasma* contamination within the biopharmaceutical industry *Biologicals* **38** 211
[90] Chatterjee R 2007 Cases of mistaken identity *Science* **315** 928
[91] Podolak E 2010 Ending cell line contamination by cutting off researchers *Biotechniques Online News*
[92] Nardone R M 2007 Eradication of cross-contaminated cell lines: a call for action *Cell Biol. Toxicol.* **23** 367
[93] Nardone R M 2008 Curbing rampant cross-contamination and misidentification of cell lines *Biotechniques* **45** 221
[94] Jornitz M W and Meltzer T H 2000 Identifying the sterilizing filter *Pharm. Technol.* **24** 38–46
[95] Khandpur P and Gogate P R 2016 Evaluation of ultrasound based sterilization approaches in terms of shelf life and quality parameters of fruit and vegetable juices *Ultrason. Sonochem.* **29** 337–53
[96] Cheung H Y 2012 Endospores as biological indicator for the validation of sterilization process *Hong Kong Pharm. J* **19** 158–64
[97] Honegger P 2001 Overview of cell and tissue culture techniques *Curr. Protoc. Pharmacol.* **4** 12.1.1-12.1.12
[98] Jones B and Stacey G 2001 Safety considerations for *in vitro* toxicology testing *Cell Culture Methods for In Vitro Toxicology* ed G N Stacey *et al* (Dordrecht: Kluwer) pp 43–66
[99] Merten O W 2002 Virus contamination of cell cultures—a biotechnology view *Cytotechnology* **39** 91–116
[100] Nakai N, Kawaguchi C, Kobayashi S, Katsuta Y and Watnabe M 2000 Detection and elimination of contaminating microorganisms in transplantable tumors and cell lines *Exp. Anim.* **49** 309–13
[101] Nicklas W, Kraft V and Meyer B 1993 Contamination of transplantable tumors, cell lines and monoclonal antibodies with rodent viruses *Lab. Anim. Sci.* **43** 296–9

[102] Bhatia S 2018 *Introduction to Pharmaceutical Biotechnology: Enzymes, proteins and bioinformatics* vol 2 (Bristol: IOP Publishing)
[103] Bhatia S 2018 *Introduction to Pharmaceutical Biotechnology: Basic Techniques and Concepts* vol 1 (Bristol: IOP Publishing)

Chapter 6

Characterization of cultured cells

6.1 Introduction

Due to the change in cellular environment, there is a remarkable difference been the characteristics of *in vivo* and *in vitro* cells. A number of unique features of the *in vitro* cell cultures are:

- The *in vitro* environment encourages proliferation and spreading of unspecialized cells (i.e. a stem cell is an unspecialized cell that is capable of replicating or self-renewing itself and developing into specialized cells of a variety of cell types).
- In comparison to *in vivo* cells, cultured cells cannot perform differentiated and specialized functions.
- The 3D architecture of cells present in an *in vivo* environment cannot exactly mimic *in vitro* conditions.
- Under *in vitro* conditions cell–cell interaction is less.
- Certain cells which do not show their normal growth pattern can be cultured normally under *in vitro* conditions.
- The regulation of cellular programming cannot be achieved to the same level as with *in vivo* conditions.
- Cellular genetic regulation cannot be programmed under *in vitro* conditions.
- The influence of physical and chemical factors on cultured cells is different than in cells present *in vivo.*

The microenvironment in which cells are cultured plays an important role in determining the growth and development of the cells. The substrate sometimes offers more resistance to the cells than they experience *in vivo*, which can ultimately affect the growth of the cells. Culturing epithelial cells on porous filters mimics *in vivo*-like conditions by allowing nutrient supply from the basolateral compartment and therefore promotes cell differentiation and closer (morphological as well as functional) similarity to the native tissue [1].

doi:10.1088/2053-2563/aafac0ch6

The resistance offered by the substrate, also called substrate stiffness, as well as the media–air interface, are considered the most important mechanical forces acting on cells that determine cell behavior. They control cellular programming, in particular signaling, which can ultimately affect the growth, survival and motility of the cultured cells. The stiffness optima (stiffness bearing capacity or elasticity) for different various adherent cells differ extensively. This variation is not constant as it changes during cell proliferation and differentiation [2]. Thus, to control cell behavior the interaction with these artificial surfaces has to be studied properly. Several factors, such as time and contact area, have to be studied properly, as mentioned in figure 6.1. It is very important to monitor the attachment and spreading of animal cells to artificial surfaces. Only a few practical approaches are available for investigating cell–substrate interactions in a quantitative way. One of the most acceptable methods is electric cell–substrate impedance sensing (ECIS) [3–5]. Small gold film electrodes (250 mm diameter) are utilized in this method. These electrodes are deposited over the bottom of cell culture dishes and determine the electrode impedance. As the cells attach and spread on the electrode surface, the impedance is altered, and the change can be used to analyze cell behavior [3–5].

Thus the substrate plays an important role in determining cells behavior. Other conditions of the microenvironment also determine the growth of the cells, such as the type of substrate, i.e. monolayer (substrate is solid) or suspension (substrate is liquid), the composition of the medium, hormones and growth factors, the composition of the gas phase and the temperature of the medium. Other parameters, such as the adhesion, proliferation, differentiation, metabolism, initiation and evolution of cells, should also be properly studied to evaluate the microenvironment.

Cultured cells derived from solid tissue always grow as adherent monolayers. Cells obtained from tissue aggregation or subcultures attach to the substrate and then start multiplying. Previously, the glass materials used for substrates were negatively charged, however, recently plastics such polystyrene treated with electric ion discharge have become common. Cell adhesion occurs when the cell itself secretes a certain protein into the extracellular matrix which later binds with the substrate, such as glass or plastic, to form a layer of proteins over the surface to encourage adhesion. It has been observed that to improve cell adhesion, the substrate material should be preconditioned with culture so that the cells secrete the proteins which help in the subsequent binding of culture cells, as shown in figure 6.2. There are certain proteins called cell adhesion proteins which are responsible for cell–cell adhesion and cell–substrate adhesion. These proteins have the tendency to

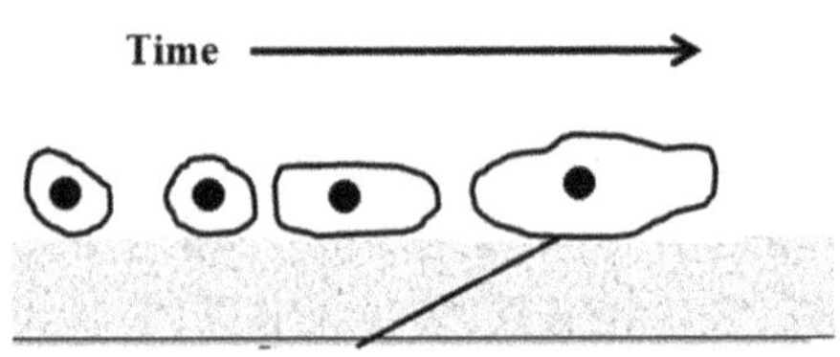

Figure 6.1. Cell spreading over substrate and the associated increase of adhesion.

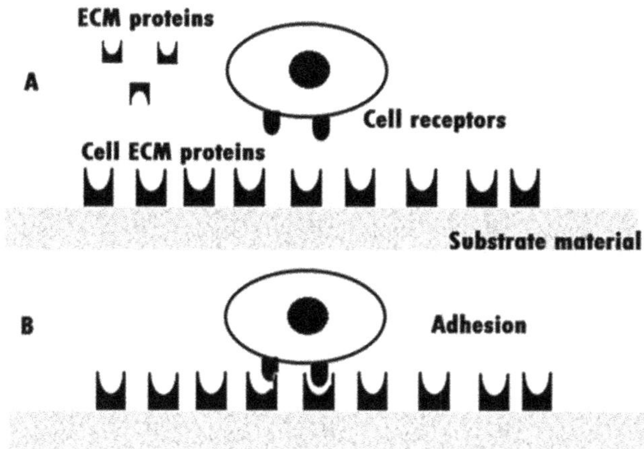

Figure 6.2. Cellular adhesion to the substrate by secretion of ECM proteins.

bind with the cytoskeleton of cultured cells and the substrate. Broadly, these proteins can be divided into three categories, calcium-dependent, calcium-independent and proteoglycans.

6.1.1 Calcium-dependent proteins (cadherins)

Calcium ions are present throughout the body as intracellular mediators of several cellular processes. Cadherins play an important role in cell–cell contact formation by cadherin–cadherin interactions. Such interactions are gated by extracellular Ca^{2+}. This interaction encourages trans-junctional interactions [6]. Cadherins are a subtype of cell–cell adhesion molecule. They are further subdivided into separate adhesive specificities and tissue distributions. Cadherin-integrated liposomes have potential applications in drug delivery systems with high potential for use as new cell-specific proteoliposomes [7, 8]. These are transmembrane glycoproteins that mediate intercellular adhesion in the presence of extracellular calcium.

6.1.2 Calcium-independent proteins

Various principles have been used to differentiate between various systems of embryonic cell adhesion. Principles such as binding specificity, dependence on Ca^{2+} and involvement of particular cell surface molecules have been used to differentiate cell adhesion systems [9]. These include transmembrane glycoproteins molecules called integrins, basically cell adhesion receptors which are responsible for cell–cell or cell–substrate interaction. These receptors work with the receptors called cadherins.

6.1.3 Proteoglycans

Generally proteoglycans are defined as transmembrane proteins that are heavily glycosylated and have low affinity to transmembrane receptors. Proteoglycans can potentially bind with collagen and growth factors. Proteoglycans belong to the family of extracellular matrix proteins. These proteins are involved in the cascade of molecular events that control cell adhesion, migration and proliferation. Various structural assortments and their distribution in tissue advocate their functional flexibility [10]. Molecular interactions and the potential to control certain molecules with proteoglycans always decide their versatility. Various investigations have proved that proteoglycans play an important role in cellular events by functioning either as receptors or as ligands for molecules [10].

6.2 Cell proliferation, differentiation and metabolism

This biological process involves four different phases: the M phase or mitotic phase (when two chromatids form daughter cells); the G1 or gap 1 phase (check whether the cell proceeds to DNA synthesis or re-enters the cell cycle); the S phase (DNA synthesis and replication) and the G2 phase (re-entry into mitosis) (figure 6.3). The maintenance of the integrity of the DNA is governed by the biological repair mechanism. In the case where DNA repair is not possible, then this is determined by two check points at the beginning of DNA synthesis and ultimately results in programmed cell death, called apoptosis.

In a culture, the cell cycle is controlled by physicochemical parameters, e.g. supplementation of growth factors in a low cell density culture facilitates cells entering into the cell cycle, however, a high cell density impedes the cell cycle and thus affects cell proliferation. In addition to environmental conditions, the cell cycle is controlled by certain intracellular factors, such as the expression of cyclins promoting and the p53 and Rb genes inhibiting the cell cycle. Low cell density, low calcium concentration and growth factors encourage cellular differentiation, whereas high cell density, high calcium concentration and supplementation of differentiation inducers can promote cell proliferation. As suggested above, contrasting conditions are required for cell proliferation and differentiation. Recently, it has been discovered that cells under an *in vitro* environment retain their originality for longer in three-dimensional environments, which can be offered by various matrices, such as cellulose, collagen gel and glycoproteins matrices. Recently, several researchers have also been working on the molecular events involved in the dedifferentiation process, which is an irreversible loss of the specialized properties of cells when they are cultured under *in vitro* conditions. It is an important stage that determines cell fate in the context of cellular adaptations in a different environment. Most of the focus of current researchers is to encourage

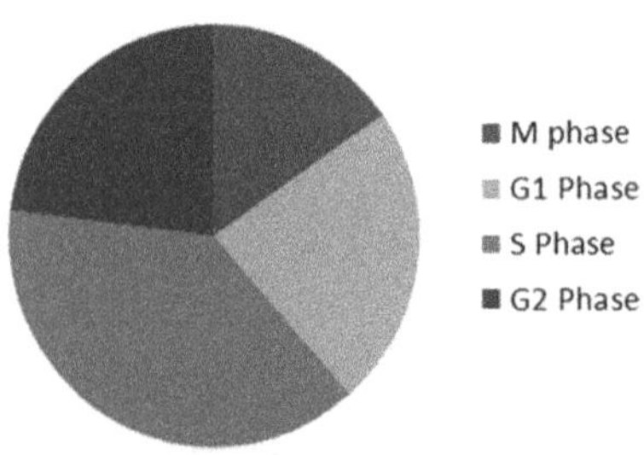

Figure 6.3. The cell cycle.

deadaptation, which is the reinduction of specialized properties lost by the cell during dedifferentiation by offering a suitable environment. Under *in vivo* conditions, stem cells develop into progenitor cells that can further produce a differentiated cell pool, whereas in an *in vitro* environment progenitor cells are predominately produced and keep on multiplying, and only a few cells out of a population can form differentiated cells. This entire process is called blocked differentiation, as demonstrated in figure 6.4.

The metabolism of cultured cells is depicted in figure 6.4. Cells under an *in vitro* environment utilize energy from glucose or glutamine to produce certain anabolic precursors, as described in figure 6.5. When glucose is degraded by the metabolic process called glycolysis, it leads to the production of lactate (due to anaerobic conditions) and pyruvate (later, a small fraction is oxidized via the Krebs cycle). Glutamine also acts as important energy sources for cultured cells. It undergoes a deamination reaction in the presence of the enzyme glutaminase to produce glutamate (enters the Krebs cycle) and ammonium ions, shown in figure 6.5.

Both glucose (glycolysis) and glutamine (deamination) are degraded by cultured cells to supply energy in the form of ATP.

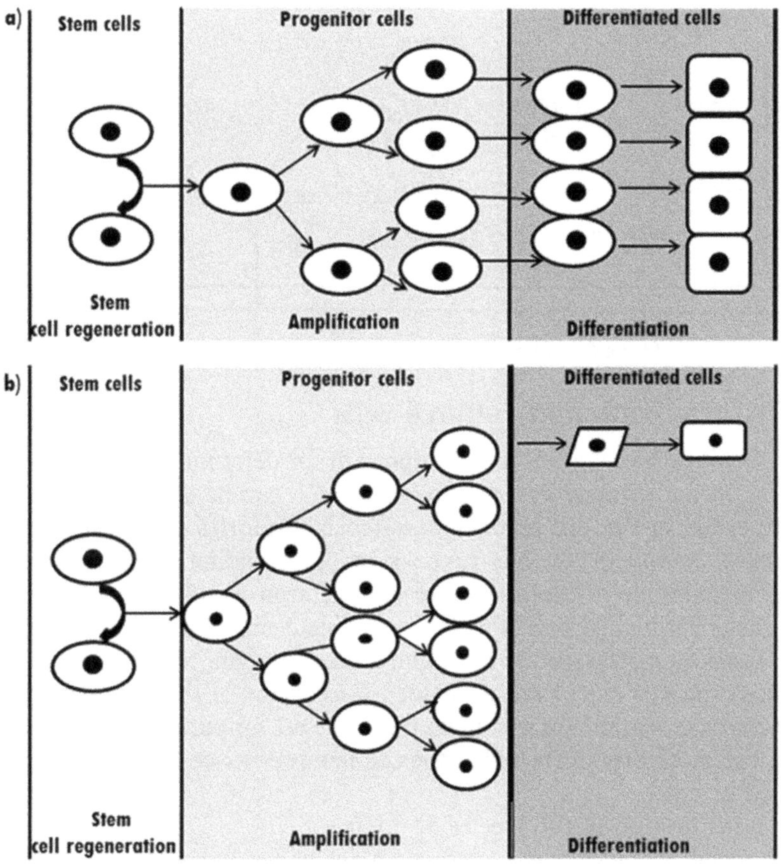

Figure 6.4. Cell differentiation: (a) differentiation of cells under *in vivo* conditions and (b) cultured cells under *in vitro* conditions (blocked differentiation).

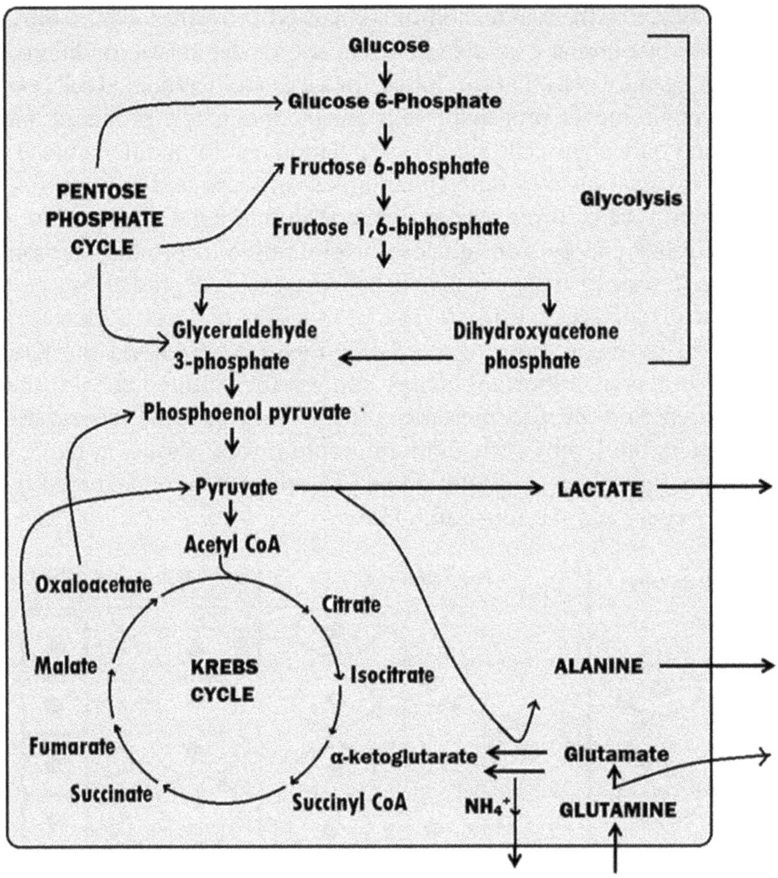

Figure 6.5. Glutamine and glucose metabolism in animal cells.

6.3 Characterization of cultured cells

The characterization of a cell line is important for determining its functionality and in proving its authenticity as a pure cell line, so it can be disseminated through cell banks to establish sharing between research laboratories and commercial companies. Special consideration has to be given to avoiding cross-contamination or misidentification due to mislabeling or confusion in handling DNA profiling. Cell line characterization involves morphological, biochemical and gene based characterization. In addition, the species of origin, tissue of origin, cell line and nature of cell line (transformed or not) have to be identified. Some of the important facts about cell line characterization are as follows (see table 6.1 for parameters and procedures):

- Close monitoring of whether the cell line is cross-contaminated or misidentified or not.
- Authentication of the species of origin.
- Association with the tissue of origin, which involves cellular identification of the lineage to which the cell belongs and also determination of the

Table 6.1. Significant parameters in cell line characterization and respective procedures.

Parameters	Procedure
Proteomics	Microarray
Karyotype	Chromosome spread with bending
Isoenzyme	Agar gel electrophoresis
Genome	Microarray
Gene expression	Microarray
DNA profile	PCR of microsatellite repeats
Cytoskeleton	Immunochemistry with antibodies
Cell surface antigen	Immunochemistry

development stage of the cells within that lineage (i.e. the stem, precursor or differentiated status).

- Determination of whether the cell line is transformed or not (i.e. whether the cell line is finite or continuous and whether the cell line expresses properties associated with malignancy).
- Determination of the factors responsible for making cell lines more vulnerable to genetic instability and phenotypic variation.
- Cell line identification within a group from the same origin (selected cell strains, or hybrid cell lines).

6.4 Parameters of characterization

6.4.1 DNA profiling or analysis of gene expression

DNA profiling offers an important technique to differentiate various sources of similar cells. For this, native tissue should be accessible. This is done to corroborate the actual identity of a cell line. DNA analysis can be utilized as an alternative approach to, or in addition to, karyotyping for certain types of cell lines [11]. This is achieved using restriction endonuclease with probes to prepare DNA for analysis. Digested genomic DNA with probes at hyper-variable regions is generated. By means of this approach cell line identity and/or cross-contamination can be determined.

6.4.2 Fluorescence *in situ* hybridization (FISH) or chromosome painting

FISH is cytogenetic tool discovered in the early 1980s, which is utilized to target specific chromosomal regions within the nucleus. These regions, when tagged with a fluorescent probe, show colored signals which can be examined by means of a fluorescent microscope. Currently, most *in situ* hybridization methods use fluorescent probes to detect DNA sequences, and a number of FISH procedures are accessible to cytogeneticists, for applications in diagnosing several chromosomal abnormalities in patients. The concept of FISH is based on *in situ* hybridization, which ultimately depends on the stability of the DNA double helix, which allows the detection of tiny chromosomal changes, mainly microdeletions, which cannot be achieved by means of classical chromosomal analysis. To identify abnormalities in chromosome number or structure, a single-stranded DNA probe is used for a known piece of DNA or chromosome segment. The sequence of interest, labeled with a fluorescent tag targeted to a single-strand DNA, is denatured in place on a microscope slide. Finally, fluorescent microscopy allows the examination of more than one probe, each of which is labeled with a different color. As described by Connor in 2008, without this technique, the process is similar to searching for a needle in a haystack, where the needle is the DNA sequence of interest and the haystack is the set of chromosomes [12]. As per Connor's demonstration, the search is easier if the investigator has a powerful 'magnet' in the form of a probe labeled with a fluorescent tag (also known as the fluorescent copy of the DNA sequence of interest). Hybridization takes place when the 'magnet' binds with the 'needle' [12]. This approach helps in differentiating the genetic material of human and mouse chromosomes, in particular potential cross-contamination.

6.4.3 Major histocompatibility complex (MHC) analysis

A major role of the MHC is to bind with peptide fragments from pathogens. Afterwards, these proteins display the same complex on the cell surface for their detection by suitable T cells. The results of such an event are almost always harmful, as during this process pathogen-infected cells are killed [13]. Moreover, this event results in the activation of macrophages to kill pathogens in their intracellular vesicles as well as the activation of B cells to synthesize antibodies that either remove or reduce the effect of extracellular pathogens. However, if the pathogen has mutated so that it avoids presentation by MHC molecules, then it would be difficult to trigger the usual immune response [13]. There are some important functionalities of the MHC which make it impossible for pathogens to avoid immune responses. One of the most important properties is the polygenic nature of the MHC, as it is made up of a number of different MHC class I and MHC class II genes, in such a way that each individual retains a set of MHC molecules with diverse ranges of peptide-binding specificities. Another important functionality of the MHC is its highly polymorphic nature, as there are several variants of every gene in the population [13]. Currently the MHC is used to characterize a number of cells that are vital determinants of β-cell destruction in type 1 diabetes [14].

The MHC is made up of the cell surface proteins necessary for the acquired immune system to identify foreign molecules. This determines the histocompatibility or tissue compatibility. Histocompatibility is the property of having similar alleles of a set of genes called the human leukocyte antigens (HLA typing). Usually, MHC analyses coupled with lineage specific markers are used to determine the immunological capability of an individual. Each individual has their own specialized proteins (antigens) present on the surface, called the HLA, which is a major part of the MHC. In fact, each nucleated cell in the body contains these proteins and the genes that code for them. Everyone has an inherited a combination of HLA protein based surface antigens present on the surface of his or her white blood cells (leukocytes) and other cells that contain a nucleus. HLA testing identifies the major HLA genes a person has inherited and their corresponding antigens that are present on the surface of their cells. Enucleated cells such as human red blood cells do not generally have easily detectable HLA antigens, although certain HLA antigens are sometimes expressed strongly enough to be detected by classical blood grouping techniques. One of the major roles of HLA antigens is in immune system management, in particular to help the body's immune system identify cells which are 'self' and those which are 'foreign' or 'non-self'. Cells which are identified as 'foreign' can stimulate an immune response through the production of antibodies. These proteins play an important part in the transplantation of tissues or organs. In conjugation with functional assays related to specific features of interest, these methods offer satisfactory information to validate a cell line as well as confirm that it is suited to the task concerned.

6.4.4 Lineage or tissue markers

To investigate and effectively treat degenerative diseases, stem cells obtained from early embryos or reprogrammed from somatic cells can be used. This approach promises a future of regenerative medicine [15]. The possibility to produce a range of differentiated cells increases the requirement for new models of early mammalian development. However, a lack of suitable cell surface markers has created an obstacle in this area. Cell surface markers play an important role in the characterization of stem cells, in particular to study differentiation and reprogramming events inside and outside the cell [15]. These are the characteristic molecules for identifying cell lineages, such as cell surface markers, mRNAs or internal proteins, which help in establishing the relationship of a particular cell line to its tissue of origin and also help in identifying and isolating stem cell populations. Under an *in vitro* environment, as cells progress they develop into a specific differentiated cell type which can be considered as a lineage [15]. During the development of the lineage they procure lineage markers exact to the lineage and distinct from markers expressed by the stem cells. Some of the enzymes can also be considered as markers for sorting cells in a culture, for example, for hepatocytes the tyrosine aminotransferase is considered as the marker enzyme because of its specific expression respective to these cells, whereas tyrosinase is for melanocytes. Some of the lineage markers are discussed in the sections below.

6.4.4.1 Cell surface antigens

Cell markers or cell surface antigens are molecules residing in the cell's plasma membrane, and serve as symbols to identify and classify cells, in particular hematopoietic cells. These molecules are unique to different cell types and each cell or group of cells carries unique combinations of markers or antigens. In addition to serving as markers they also have uses such as helping in diagnosing diseases or direct treatment by identifying which molecules are present. Moreover, they are also used in the selection of epithelium from mesenchymally derived stroma with antibodies such as anti- and anti-HMFG 1, distinguishing among epithelial lineages, and identifying neuroectodermally derived cells (e.g. with anti-A2B5) [16]. Antigens of cultured cells can be utilized for the detection of the tissues or cell of origin. Recently, many antibodies have been developed for the identification of cell lines using cell surface antigens, e.g. antigens such as integrins and α-fetoprotein can be used for the identification of the cell line. Current commercially available antibodies (table 6.2) are tagged with fluorescent probes to distinguish desirable cells from a mixed population, shown in figure 6.6.

Cell adhesive interactions play important roles during many normal physiological processes, such as embryonic development and wound repair, and also during the progression of diseases such as cancer. Cell adhesion is mediated by the specific interactions of cell surface receptors with extracellular glycoproteins [30]. Our current understanding of the structure and function of cell adhesion molecules (i.e. integrins, cadherins) has affected the design and development of drugs (i.e. peptide, proteins) (figure 6.6). This impact resulted in the development of potential treatments for cancer and heart and autoimmune disorders [28].

6.4.4.2 Intermediate filament proteins

Intermediate filaments are proteins which are primordial components of the cytoskeleton and they are among the most widely used lineage or tissue markers. Glial fibrillary acidic protein (GFAP) for astrocytes and desmin for muscle are the most specific, whereas cytokeratin marks epithelial cells and mesothelium [31–33]. GFAP belongs to the family of intermediate filament proteins and is considered as the main protein component of glial intermediate filaments in differentiated fibrous and protoplasmic astrocytes of the central nervous system (CNS). These proteins are present in the neural crest and have a important role in defining and maintaining the shape of the astrocyte. This principal marker for brain astrocytes is a potential marker for several tumors with cartilaginous differentiation [31–33]. CNS injury that results in gliosis and then upregulates GFAP makes GFAP a striking candidate biomarker for brain injury screening. The brain-specific astroglial protein GFAP is a blood biomarker candidate indicative of intracerebral hemorrhage in patients with symptoms suggestive of acute stroke [31–33].

6.4.4.3 Specific cell type markers

Cell markers are used to differentiate between different cell types. Certain cells and their precursors express unique sets of proteins, such as transcription factors, enzymes, cytoskeletal proteins and receptors. Some of the examples are mentioned in table 6.3. Like all differentiation markers, they depend on the complete expression of the differentiated phenotype. Under *in vitro* conditions it is possible to maintain the transport of nutrients, as is usually done in the case of epithelial and endothelial cells using Boyden chambers where the movement of the cell or organism in response to a chemical stimulus is checked (figure 6.7).

6.4.4.4 Enzymes as markers

Enzymes are proteins that act as catalysts within cells. Due to their significant properties, they are widely used in medical diagnosis, in particular as markers.

Table 6.2. List of commercially available antibodies.

Cell type	Antibody	Role	Ref.
Epithelium	Cytokeratin	• A tumor marker for epithelial malignancies. • A noninvasive, cheap and reliable tool for efficient management.	[17]
Epithelium	Epithelial membrane antigen	• Immunohistochemical profiling in epithelial, mesenchymal and hematopoietic neoplasms. • A marker for establishing the epithelial nature of neoplastic cells. • Glycoproteins isolated from human milk fat globule membranes.	[18]
Hepatocytes	Albumin	• Hepatitis B surface antigen is known to bind to human serum albumin polymerized by glutaraldehyde. • Binds irreversible to cysteine(s) of the small HBs protein.	[19]
Breast epithelium	α-lactalbumin	• The major whey protein of human milk. • Possess some important functions of immunologic defense and can induce apoptosis in transformed embryonic and lymphoid cell lines. • A complex in human milk that induces apoptosis in tumor cells but spares healthy cells.	[20]
Colorectal adenocarcinoma	Carcinoembryonic antigen	• To help diagnose and manage certain types of cancers.	[21]
Lung adenocarcinoma	Carcinoembryonic antigen	• Cell surface glycoprotein that functions as an intracellular adhesion molecule. • Because of its high expression in cancer cells and secretion into serum, CEA has been widely used as a serum tumor marker.	[21]
Prostatic epithelium	Prostate specific antigen	• Diagnosis and treatment of prostate cancer. • To detect prostate tumors while they are still small, low-grade and localized.	[22]

(Continued)

T cells	Intracellular cell adhesion molecule	• This cell adhesion molecule is expressed by several cell types, including leukocytes and endothelial cells. • It can be induced in a cell-specific manner by several cytokines, for example, tumor necrosis factor-alpha, interleukin-1 and interferon-gamma, and inhibited by glucocorticoids.	[23]
Fetal hepatocytes	α-fetoprotein	• A plasma protein produced by the embryonic yolk sac and the fetal liver. • In serum, amniotic fluid and urine functions as a screening test for congenital disabilities, chromosomal abnormalities, as well as some adult occurring tumors and pathologies. • This tumor marker is a glycoprotein encoded by the AFP gene on chromosome 4q25.	[24]
Placental epithelium	Human chorionic gonadotrophin	• Glycoprotein hormone consisting of two noncovalently bonded subunits, alpha and beta. • Placental hormone that stimulates secretion of the pregnancy-sustaining steroid progesterone.	[25]
Anterior pituitary	Human growth hormone	• Transmembrane proteins that includes the prolactin receptor and a number of cytokine receptors.	[26]
Mesodermal cells	Vimentin	• Intermediate filament protein which is characteristically upregulated in cells undergoing the epithelial-to-mesenchymal transition, which is a critical event in the induction of cell motility and increases survival, both under physiological situations such as wound healing and during development, as well as in malignant cells undergoing invasion and metastasis.	[27]
All cells	Integrins	• Cell adhesion receptors. A transmembrane glycoprotein that binds effectively with fibronectin, the $\alpha 5,\beta 1$ integrin is the major fibronectin receptor on most cells. This integrin mediates such cellular responses to fibronectin substrates as adhesion,	[28]

		migration, assembly of extracellular matrix and signal transduction. • Integrin ligands, such as fibronectin, are not passive adhesive molecules but are active participants in the cell adhesive process that leads to signal transduction. • Integrins helps in diagnosis and treatment of human disease.	
All cells	Actin	• The actin cytoskeleton plays an important role in many cellular processes, including cell motility, cytokinesis and intracellular transport.	[29]

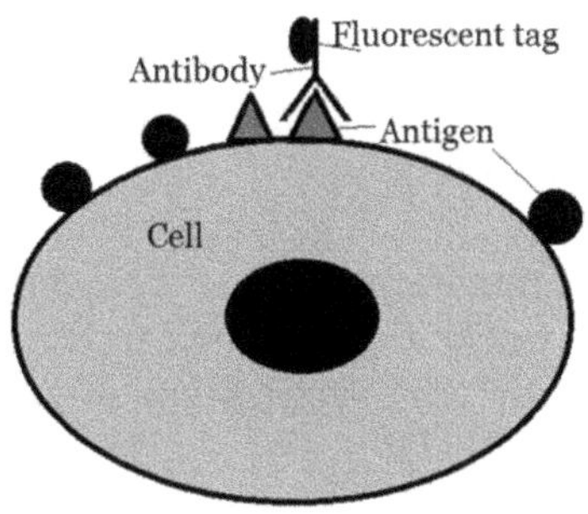

Figure 6.6. Cell surface antigens.

Table 6.3. Specific cell type markers.

Markers	Cells
Hemoglobin	Erythroid cells
Myosin or tropomyosin	Muscle
Melanin	Melanocytes
Serum albumin	Hepatocytes

Several reports are available where enzymes are considered as endoplasmic reticulum markers, cardiac markers and diagnostic markers [34–38]. Several enzymes are considered as markers in the diagnosis of diseases such as cancer, neurodegenerative disorders, etc. Enzyme markers can be studied through blood investigations that examine specific enzyme activity in the human or any specific biological model body. Defects that are passed from generation to generation

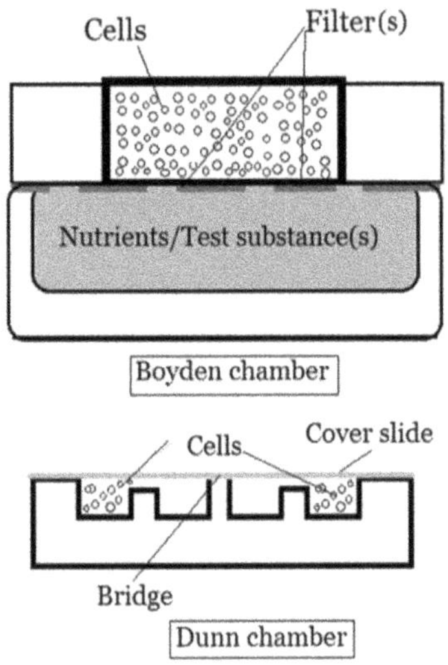

Figure 6.7. Transport of nutrients using different chambers.

usually affect enzyme structure and function. This can often lead to abnormality. Some enzymes are affected by several genes. Enzymes can be identified in cultured cells using three parameters, as follows.

The constitutive level. Constitutive enzymes are those enzymes which are produced in constant amounts without regard to physiological demands or the concentration of the substrate. In this case the characterization of enzymes is performed in the absence of inducers or repressors [39].

Induced or adaptive level. An adaptive enzyme or inducible enzyme is an enzyme that is expressed only under favorable conditions, i.e. in response to inducers and repressors [39].

Isoenzyme polymorphisms. Enzymes that are different in their amino acid sequence or composition but nevertheless catalyze similar chemical reactions are called isoenzymes. They have a number of advantages over traditional morphological characteristics as they are not affected by environmental factors, making identification easy in the early stages of development and thereby saving both time and space [42]. Initially, enzyme polymorphisms were investigated in experimental populations of *Drosophila melanogaster* [40]

Isoenzyme variability is an extensive source of genetic markers that can be utilized for the identification of any organism or hybrid, monitoring of genetic diversity and quantification of genetic relationships among populations [41]. Variations among these enzymes are called isoenzyme polymorphisms, which were first studied using isoenzymes of the almond [43]. They detected variation in six different enzymes:

- 6-phosphogluconatedehydrogenase.
- Aspartate aminotransferase.
- Glucose phosphate isomerase.
- Leucine aminopeptidase.
- Malate dehydrogenase.
- Phosphoglucomutase.

These markers can also be utilized for the identification of cell lines. Some enzyme markers for cell line identification are mentioned in table 6.4.

6.4.4.5 Regulatory factors as markers

The growth and development of differentiated cells is usually controlled by several environmental factors, such as nutrients, hormones, the matrix and adjacent cells. Thus the supplementation of regulatory markers such as hormones (e.g. hydrocortisone) or specific growth factors, or the growth of cells on extracellular matrix of the correct type, can be used to identify the type of cell and to monitor growth and development [54]. It has been discovered that epidermal growth factor plays an important role in tumor progression and could be useful as potential metastatic and/or prognostic marker for gastric carcinoma [53]. Similarly, glycoprotein markers, such as vascular endothelial growth factor, have been considered as potent mediators of both blood vessel and lymphatic vessel formation in the context of tumor biology. Based on recent reports, it has been discovered that the growth factor Ki-67 may potentially mark therapeutic targets in the treatment of osteosarcoma [55].

6.4.4.6 Lineage originality

Lineage markers are tissue or cell-specific markers, and they are often more characteristic of the function of the cell than its embryonic origin. These markers determine the fidelity of cell lines after their characterization or authentication, which helps in sorting cells.

Table 6.4. Enzyme markers for cell line identification.

Enzyme markers	Cell	Ref.
Alkaline phosphatase	Enterocytes	[44]
Angiotensin converting enzyme	Endothelium	[45]
Creatine kinase	Muscle cells	[46]
Creatine kinase	Neurons	[47]
Glutamyl synthase	Brain cells	[48]
Nonspecific esterase	Macrophages	[49]
Sucrase	Enterocytes	[50]
Tyrosinase	Melanocytes	[51]
Tyrosinase	Hepatocytes	[52]

6.4.5 Unique markers

There is a class of markers that is utilized to check whether a cell line is authentic or not. These are the markers or antigens which are unique to different cell types (table 6.5). It can be used to distinguish between cell lines from the same tissues but different donors.

6.4.6 Transformation

Transformed cell lines offer the benefit of nearly unlimited obtainability, however, they have the drawback of retaining very few of the original *in vivo* characteristics. Thus their characterization is essential as there are always substantial differences between primary and transformed cultures. Characterization of transformed cell lines is important to detect the variations from the first culture. Characterization of transformed cells can be achieved using various approaches as discussed in the follow sections.

6.4.6.1 Cell morphology

Morphological characterization is a classical, simple and direct approach for identifying cells. Table 6.6 describes the different morphological characteristics that can be observed. However, morphological characterization needs to be carried out carefully as morphology is dependent on the culture environment, but can ultimately help in determining the plasticity of the cells, i.e. the ability of the cells to convert from one cell type to another or the ease of transition between these cell fates. Thus morphological characterization helps in discovering the plasticity of cellular morphology in response to different culture conditions. Based on the morphological characteristics, cells are divided into various categories such as fibroblastic or fibroblastoid (fibroblast-like), epithelioid (epithelial-like), lymphoblast-like, endothelial, neuronal, etc. To understand this more clearly, we can take the example of epithelial cells multiplying and developing in the epicenter of a confluent sheet. These cells are typically regular and polygonal, with a clearly defined birefringent edge. However, similar cells, when growing at the edge of a

Table 6.5. Unique markers for different cell types.

Markers	Use to identify...
Specific markers for chromosomal abnormalities [56]	Deletions, translocations, polysomy
Major histocompatibility [57]	Group antigens (e.g. HLA in humans), which are highly polymorphic
DNA fingerprinting or SLTR DNA profiling	Specific cell lines [59]
Enzyme deficiencies	Thymidine kinase deficiency [58]
Drug resistance such as vinblastine resistance	Expression of P-glycoprotein [60]

Table 6.6. Morphological characterization of different cells.

Cells	Morphological characterization	Morphological differences
Fibroblastic cells (or fibroblast-like)	These are bipolar or multipolar and have an elongated shape, with the length usually more than twice the width, and grow attached to a substrate.	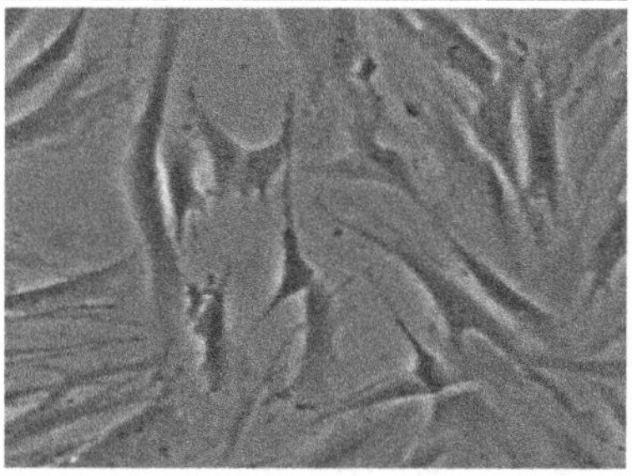
Endothelial cells	These are very flat, have a central nucleus, are about 1–2 μm thick and 10–20 μm in diameter. Monolayer cells that are polygonal with more regular dimensions, and that grow in a discrete patch with other cells.	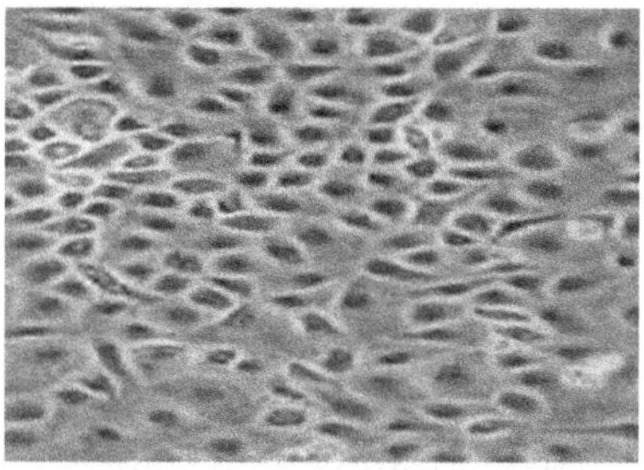
Neuronal cell line	These occur in diverse shapes and sizes, but can roughly be divided into two basic morphological classes: • Type I—with long axons used to move signals over long distances. • Type II—without axons.	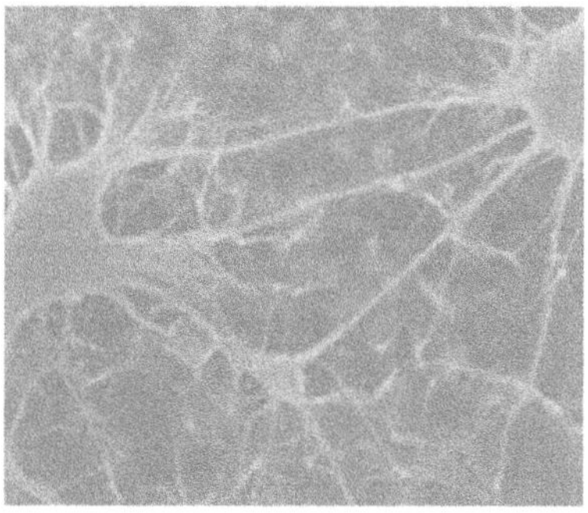
Lymphoblast-like cells	These are spherical in shape and typically grown in suspension without attaching to a surface.	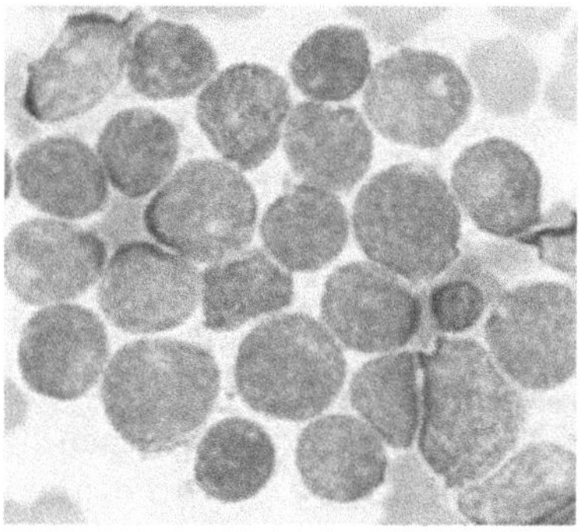

patch, form irregular and distended structures. These structures can detach from the patch to form fibroblast-like structures. Changes in the substrate and the constituents of the medium can also affect cellular morphology. Comparative observations of cells should always be made at the same stage of growth and cell density in the same medium, and for growth on the same substrate. Precursor cells that are still capable of dividing are called blast cells, for example: a fibroblast is a proliferative precursor of a fibrocyte; a myoblast is a proliferative precursor of a myocyte; and a lymphoblast is a proliferative precursor of a lymphocyte. When the identity of the cells has not been confirmed, the terms 'fibroblast-like' (or 'fibroblastoid') and 'epithelium-like' (or 'epithelioid') should be used.

6.4.6.2 Microscopy

Usually, inverted microscopy is used for the characterization of cells, the principles of which depend on a source of transmitted light and a condenser positioned on top of the stage, as shown in figures 6.8(a) and (b). This position allows the transmission of light in an inverted direction. In comparison to compound microscopy, where objectives are placed above the stage, in this case objectives are positioned below the stage and thus the specimen is studied from the bottom of the cell culture vessel.

Inverted microscopy is a significant technique in the animal tissue culture laboratory; however, it is frequently used incorrectly. This tool is important in characterization as it allows the entry of light into the culture vessel from below. Usually, the thickness of the closed culture vessel makes observation problematic, thus in inverted microscopy the culture vessel is positioned on the stage, irradiated from above and observed from below. The large width of the vessel restricts the working distance and objective magnification is usually restricted up to 40×, thus the utilization of phase-contrast optics is encouraged. In phase-contrast microscopy the pathway of the illuminated source of light is obstructed by a dark ring in the objective. After the placement of the specimen only diffracted light is visible, which allows unstained cells to be observed with higher contrast than is available by normal illumination. By using different types of microscopy techniques it is feasible to maintain a record of photographs at periodic intervals to check the morphological differences of cells at different developmental stages.

6.4.6.3 Staining

Staining in the context of characterizing cell lines is used during chromosomal analysis (in chromosomal banding and painting). A polychromatic blood stain stains polychromatic cells produced during polychromasia, a disorder where the body prematurely releases immature red blood cells from the bone marrow in unusually high numbers into the bloodstream [61]. These red blood cells have high affinity for basic stains, in contrast to the commonly utilized acidic stains. These cells show a dark blue or gray color when stained with basic stains such as Giemsa [62]. This type of stain offers a way to distinguish premature cells from normal cells. This stain is usually mixed with May–Grunwald stain for blood staining purposes, however, not when staining cultured cells. Giemsa alone can stain the nucleus pink or magenta, the nucleoli dark blue and the cytoplasm pale gray-blue. Additionally, it also stains

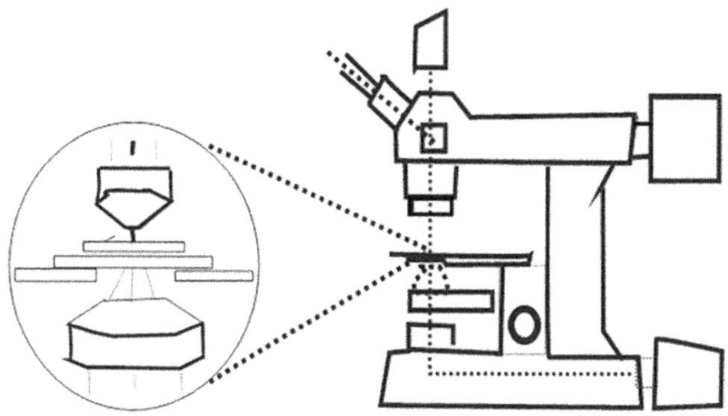

(a)

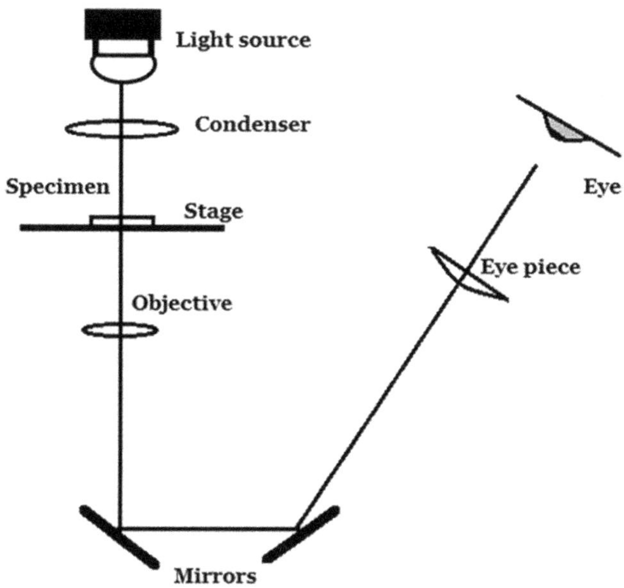

(b)

Figure 6.8. Inverted microscopy: (a) the microscope and (b) the light path.

alcohol or formaldehyde fixed cells, however, it will not work properly if the sample is totally anhydrous [63].

6.4.6.4 Karyotype

Chromosome content or karyotype is a more reliable method for identification of a cell line and its respective species, sex and origin. This approach can effectively differentiate normal and transformed cells as the chromosome number is more stable in normal cells (excluding mice, where the chromosome content of normal cells can change quickly after transfer into culture).

6.4.6.5 Chromosome banding

Certain dyes stain certain regions of the chromosomes more intensely then others, producing characteristic banding patterns which are unique for each individual chromosome [64, 65]. Chromosome banding is chromosomal treatment to display the typical arrangements of horizontal bands. This imparts a distinctive appearance to each banding pattern, which helps in the identification of desirable patterns. Banding also allows the identification of chromosome deletions, duplications and other types of structural rearrangements. The banding pattern is characteristic for each chromosome pair [64, 65]. Different types of banding patterns exist, such as those mentioned in tables 6.7 and 6.8. The chromosome banding procedure is illustrated in figure 6.9.

These approaches are used to allow individual chromosome pairs to be identified when there are slight morphological differences between them. There are certain other approaches which can be utilized, such as rather than using trypsinization for protein digestion, trypsin and EDTA can be used in combination [64, 65].

6.4.6.6 Chromosome painting

Chromosome paints are accessible on the market for different purposes, however, the hybridization and identification procedures differ for each commercial source [69–72]. There are different types of chromosomal painting, as listed in table 6.9. As discussed above, karyotypic examination is usually performed using chromosome banding, with dyes that differentially stain the chromosomes. Thus after staining, each chromosome is identified by its unique banding pattern. However, these classical banding procedures cannot distinguish several complex chromosomal abnormalities [69–72]. Thus new approaches for karyotyping are required. These new procedures, such as spectral karyotyping (SKY) and multicolor fluorescence *in situ* hybridization (M-FISH), allow the instantaneous visualization of all 23 human chromosomes in different colors. This technique allows the interpretation of a particular chromosome or chromosome segment which can be identified by DNA hybridization (*in situ* hybridization technology) using numerous fluorescence-labeled DNA fragments from the full length of a chromosome or segment [69–72]. This allows the extra chromosomal and cytoplasmic localization of specific nucleic acid sequences, such as specific mRNA species.

6.4.6.7 Chromosome analysis

Chromosomal analysis involves chromosome counting and karyotyping, which allows species identification of the cells. However, karyotyping is tedious, thus chromosome counting, with a quick assessment of complete chromosome morphology, may be necessary to identify or remove a suspected cross-contamination. The following procedures have been utilized for analyzing chromosome matching.

Chromosome counting. There are certain classical rules for counting the number of chromosomes, such as that the number of functional centromeres is always equivalent to the number of chromosomes. Similarly, the number of DNA molecules is always equivalent to the number of chromatids.

Table 6.7. Types of chromosomal banding.

Methods	Procedure
G-banding [66]	• Staining a metaphase chromosome with Giemsa stain is called G-banding. • Favorably stains particular areas that are rich in adenine and thymine to appear dark. • Chromosomal proteins are partially digested by crude trypsin, producing a banded appearance on subsequent staining.
C-banding [67]	• Staining centromeric regions and other regions containing constitutive heterochromatin. • Techniques have been developed for discriminating between human and mouse chromosomes, principally to aid the karyotypic analysis of human–mouse hybrids. These methods include fluorescent staining with Hoechst 33258, which causes mouse centromeres to fluoresce more brightly than human centromeres.
Q-banding [68]	• The staining reagent quinacrine mustard, a fluorescent stain and an alkylating agent, was the first chemical to be used for chromosome banding. • Quinacrine bright bands are composed primarily of DNA rich in bases adenine and thymine. • Trypsinization is not required for quinacrine banding. • Stains the cells in 5% (w/v) quinacrine dihydrochloride and 45% acetic acid. This is followed by rinsing the slide and mounting it in deionized water at pH 4.5

Table 6.8. Simplified description of chromosomal banding.

Technique	Procedure	Banding pattern
G-banding	• Proteolysis with trypsin. • Staining with Giemsa dye.	Dark bands A–T rich Light bands G–C rich
R-banding	• Heat denaturing. • Staining with Giemsa dye.	Dark bands G–C rich Light bands A–T rich
Q-banding	• Staining with quinacrine mustard dye.	Dark bands A–T rich Light bands G–C rich
C-banding	• Denaturing with barium chloride. • Staining with Giemsa dye.	Dark bands represent constitutive heterochromatin

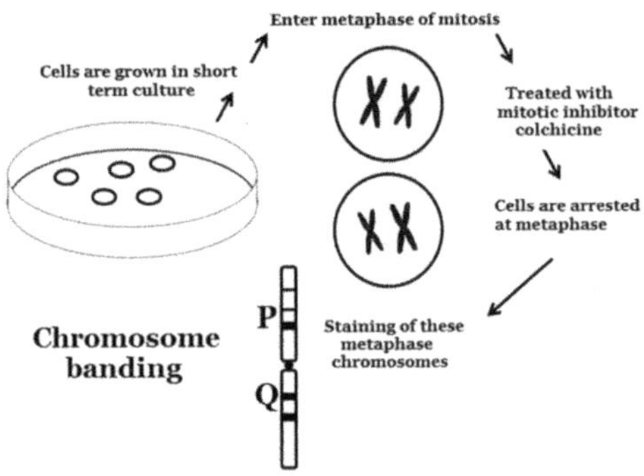

Figure 6.9. The procedure involved in chromosomal banding.

Table 6.9. Types of chromosomal painting.

Technique	Details
Specifically spectral karyotyping (SKY) [73–75]	• A potential whole-chromosome painting technique that facilitates the concurrent imagining of each chromosome in various colors. • Five spectrally different dyes are used in combination to create a mixture of probes distinctive to each chromosome. • The probe mixture is hybridized to metaphase chromosomes on a slide and is then detected with a spectral interferogram cube, which facilitates the extent of the entire emission spectrum with a single exposure. • The image is sorted out by a computational tool which can differentiate differences in color not visible to the naked eye by allocating a numerical value. • SKY can distinguish chromosomal material of unknown origin, complex rearrangements, translocations, large deletions, duplications and aneuploidy. • SKY retains several disadvantages, such as the inefficient detection of microdeletions and inversions. • Another disadvantage is that it can only be performed on dividing cells.
Multicolor fluorescence *in situ* hybridization (M-FISH) [76, 77]	• Based on chromosome painting. • M-FISH identifies translocations and insertions. • A reliable tool for diagnostic applications, and interphase nuclei can be hybridized with the FISH probe. • M-FISH is a filter based technology which does not depend on specialized instrumentation for its implementation (in contrast to SKY).

Karyotyping. Karyotyping involves systematic and ordered representation of the entire chromosome of a cell including:

- the appearance of chromosomes in the nucleus of a eukaryotic cells
- the total number of chromosomes.

A karyotype is an orderly display of magnified images of an individual's chromosomes that are presented by arranging chromosomes of somatic complement in descending order of size keeping their centromeres in a straight line:

- Longest chromosome on extreme left.
- Shortest chromosome on right.
- Sex chromosomes (allosomes) on extreme right.

6.4.6.8 DNA analysis

There are several methods available for the analysis of DNA content:

- DNA hybridization.
- DNA fingerprinting.
- DNA profiling.

One of the methods utilized most often is the measurement of DNA using propidium iodide fluorescence with a CCD camera or flow cytometry. However, the sample preparation, in particular production of the required single-cell suspension, will destroy the structure of the specimen. DNA can be assessed in homogenates with blue fluorescent dyes used to stained DNA, these are counterstains that emit blue fluorescence when bound to dsDNA Hoechst 33258. Other DNA fluorochromes, such as DAPI (4′,6-diamidino-2-phenylindole), propidium iodide or Pico Green (molecular probes), can also be utilized to estimate DNA in the sample of interest. DAPI is a fluorescent stain that binds strongly to A–T rich regions in DNA. Estimation of DNA content is mainly useful in the characterization of transformed cells that are often aneuploid and heteroploid. Examination of DNA content is mainly utilized in the characterization of transformed cells, which are often aneuploid and heteroploid.

6.4.6.9 DNA hybridization

DNA hybridization is the procedure of forming a non-covalent, sequence-specific interaction between two or more complementary strands of nucleic acids to form a hybrid. Hybridization between specific molecular probes and unique DNA sequences by Southern blotting can offer data on species-specific regions, regions that are amplified, or altered base sequences that are characteristic to that cell line. By means of this approach strain-specific gene amplification can be encouraged, as described in table 6.10.

DNA abnormalities can be identified in restricted DNA fragments using polymerase chain reaction with a primer to form a hybrid with the sequence of interest, which allows detection in comparatively small numbers of cells. Alternatively, using *in situ* hybridization specific probes can be utilized to identify specific DNA

Table 6.10. Application of DNA hybridization.

Amplification procedure	Host cell
Amplification of the dihydrofolate reductase gene	Methotrexate-resistant cells
Amplification of the MDR gene	Vinblastine-resistant cells
Overexpression of a specific oncogene, or oncogenes	Transformed cell lines
Deletion, or loss, of heterozygosity	Suppressor genes

sequences. This approach has the advantage of exhibiting structural variations and heterogeneity within a cell population. So, overall, DNA hybridization offers information about:

- Species-specific regions.
- Amplified regions of the DNA.
- Altered base sequences that are characteristic to that cell line.

6.4.6.10 DNA fingerprinting

In the research laboratory, a DNA sample is analyzed periodically to detect any changes. Usually, cell lines from the same original DNA fingerprints remain stable in culture, i.e. retain the same or very similar DNA fingerprints even after maintaining the sample for several years [78–80]. DNA fingerprinting is a potential technique to define the origin of a cell line. If the original cell line sample or DNA from it has been retained, it can be compared against the current cell line. If they have the same arrangements of bands, the sample is a more stable sample, otherwise the sample is described as unstable (as the sample cannot withstand different external factors). In this technique, the sample is first digested with restriction enzyme to allow the splicing of bulky genetic material into small fragments [78–80]. Based on their size, DNA fragments are separated using a technique called gel electrophoresis, followed by staining, autoradiography, hybridization or fluorescent probing to increase the visibility of the sample over a nitrocellulose sheet (figure 6.10). This DNA fingerprint helps in determining the origin of a cell line. Monitoring DNA fingerprints helps in determining sample variation from the primary culture to several subcultures. Additionally, it also helps in determining cross-contamination and misidentification.

6.4.6.11 Antigenic markers

Antigenic markers, also called cell surface antigen antibodies or cell markers, are molecules or antigens that act as markers to help identify and classify cells. These markers have remarkable applications in immunostaining and ELISA assays, which can be utilized for cell line characterization with a large number of antibodies [81]. It is essential to be certain of the specificity of an antibody by using appropriate control materials. This is true for monoclonal antibodies and polyclonal antiserums alike; a monoclonal antibody is highly specific to a particular epitope. These antigenic markers include a broad class called clusters of differentiation (CD) proteins. These

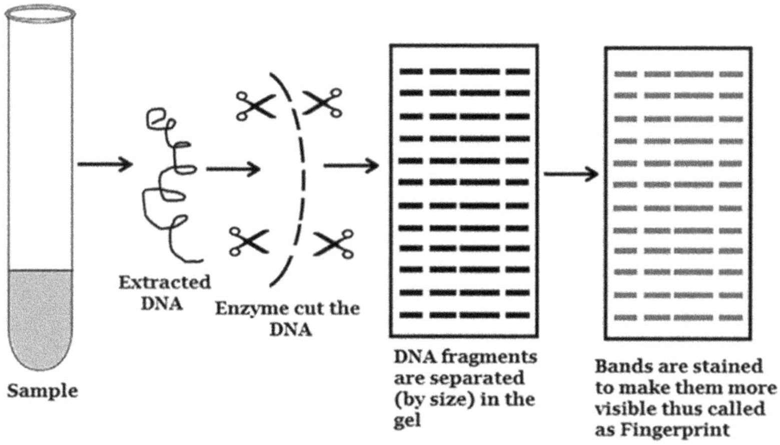

Figure 6.10. Procedure for fingerprinting.

molecules are used for the identification and study of cell surface molecules which can offer sites for immunophenotyping of cells. To isolate a particular cell type from a heterogeneous population, the unique properties of that cell type need to be exploited. Cell surface markers allow the specific binding of surface antigens to either antibodies or aptamers, and can selectively help in sorting or identifying cells of a specific surface phenotype. These cells are then identified with the help of probes such as fluorochromes and magnetic particles with which the antibodies/aptamers are labeled.

6.4.6.12 Immunostaining

Using immunochemistry it is now possible to visualize the localization and distribution of antigens in tissue sections by using labeled antibodies as specific reagents through antigen–antibody interactions. The localization of antibodies in the interior or exterior of the cell can be determined by fluorescence. During this process the antibodies are allowed to conjugate with fluorochrome, such as fluorescein or rhodamine (figure 6.11). This is usually done by the deposition of a precipitated product from the activity of horseradish peroxidase or alkaline phosphatase conjugated to the antibody. Different methods have been utilized to increase the sensitivity of detection of these approaches, mainly the *peroxidase* linked approaches.

This approach is the more sensitive and popular method. In the peroxidase–anti-peroxidase (PAP) technique, a further amplifying tier is added by reaction with peroxidase conjugated to anti-peroxidase antibodies from the same species as the primary antibody (figure 6.12) [82]. This approach was first carried out using the sandwich method, which involves three main stages:

- Application of primary antibody.
- Application of secondary antibody.
- Application of PAP complex.

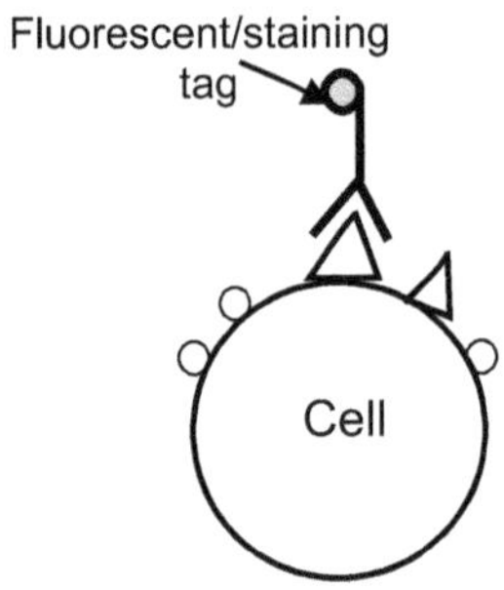

Figure 6.11. Fluorescent staining.

Sandwich procedures are still widely utilized today, with the peroxidase–anti-peroxidase procedure the most popular method. Peroxidase–anti-peroxidase is an antibody–HRP complex which causes a surge in the amount of enzyme (HRP) which can be utilized to detect a primary antibody. This gives it greater sensitivity than the direct and indirect approaches and it is used widely in diagnostic immunohistochemistry. Even greater sensitivity has been obtained by using a biotin-conjugated second antibody with streptavidin conjugated to peroxidase, alkaline phosphatase or a gold-conjugated second antibody with subsequent silver intensification.

6.4.6.13 Markers used for differentiation

Antigenic markers or enzyme activities can also be considered as markers of differentiation. These markers help in correlating cell lines with their tissue of origin. Additionally these markers define their phenotypic status. Even though they are occasionally constitutively expressed (e.g. melanin in B16 melanoma or factor VIII in endothelial cells), the expression of differentiated lineage markers may need to be induced before detection is possible.

6.5 Characterization of stem cells

Human embryonic stem cells can supply sufficient quantities of cells for transplantation and other cell replacement therapies. Their karyotypic and phenotypic stability, even after one year of culture maintenance, makes them suitable for different cell therapies. So far, twenty six human embryonic stem cell lines have been studied. Due to the remarkable similarities in the expression of their markers (similar expression profiles of surface markers), advanced characterization tools are required for their analysis. Immunochemistry based immunological techniques, in particular florescent microscopy and staining techniques, are extensively utilized for the characterization of stem cells (figure 6.13). Immunocytochemistry is a valuable tool that is utilized to evaluate the existence of a specific protein or antigen in cells, i.e. this technique helps in assessing the sub-cellular localization of proteins. This is done by means of specific antibody binding which facilitates imagining of the protein under a microscope. This allows identification of the three germ layers in *in vitro* derived embryoid bodies and detection of differentiation markers in *in vivo*

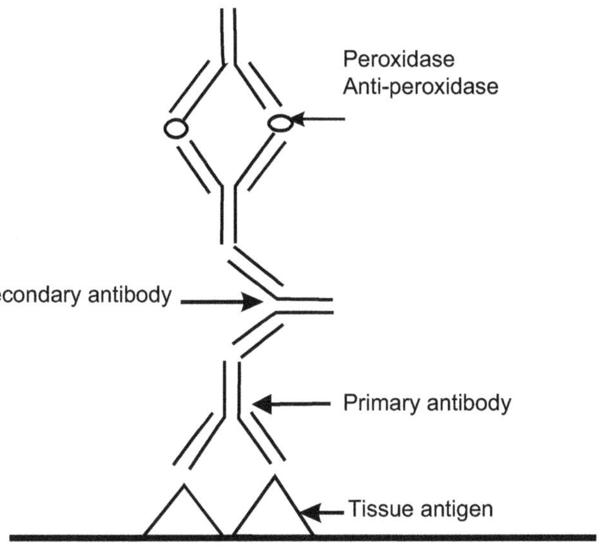

Figure 6.12. Peroxidase–anti-peroxidase complex.

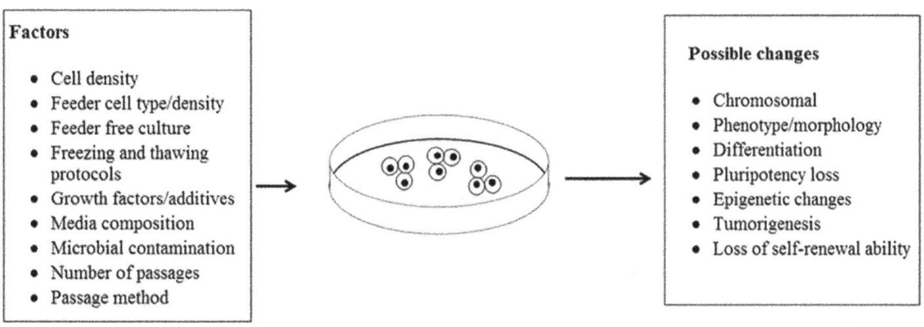

Figure 6.13. Characterization of stem cells.

generated teratomas. Characterization allows the identification of cell-specific and cytoplasmic proteins. Cell surface proteins, in particular markers of cellular differentiation (CD cells), largely act on immune cells, and can reveal information about the differentiation and development. These cell surface proteins can be used as markers of epithelial cells. In addition, the cytoplasmic proteins of epithelial cells can also be utilized for their identification. These markers can thus be utilized to characterize stem cells. The markers not only confirm the stability of stem cell lines, but can also be utilized to screen the cells during their reprogramming into induced pluripotent stem cells or to track the development of stem cell differentiation. Characterization of embryonic stem cells and induced pluripotent stem cells can be achieved using pluripotency protein or RNA marker kits and stem cell gene arrays.

Stem cell cultures should be routinely monitored for any changes in the morphology, number and appearance of chromosomes in the nucleus and the cells' capability to differentiate. When cells start adapting to a culture after experiencing transition from an *in vivo* to *in vitro* environment, certain changes take place in the culture. These changes will persist when cultures are maintained in the form of cell lines. Moreover, at this stage cells show their differences in the rate of proliferation. Some cells grow fast and proliferate more rapidly (dominant or rapidly growing cells) than others which will lead to a change in the composition of stem cell populations in culture. A population of cells which contains two or more populations of cells with different genotypes is called a mosaic population. Such a population needs to be characterized properly before its utilization. There are certain cell populations of stem cells, such as embryonic stem cells and induced pluripotent stem cells, which are more vulnerable to genetic instability must be carefully monitored for chromosomal changes. Several techniques are available for genetic characterization of stem cells, as described in the following sections.

6.5.1 Comparative genomic hybridization

Comparative genomic hybridization (CGH), involves the comparison of two different types of genomes, for example a comparison of a normal cell genome with a tumor cell of the same origin. The CGH tool can be utilized to understand the difference between two distinct cells from the same origin at the molecular, i.e. chromosomal, level [83–86]. To study this process in more detail, at the metaphasic stage, cells are treated with an alkali to denature the DNA. This will help in deriving single-strand DNA. Both samples are tagged independently with fluorescent dye. After this, hybridization is encouraged by mixing the samples. From fluorescence analysis using a suitable fluorescent detector, both samples can be analyzed simultaneously to generate a plot [83–86]. This plot helps in examining the extent of similarities and differences between the tumor and normal genomic material. This plot gives an idea of the ratio of fluorescence derived from normal and tumor sample, e.g. F1 (fluorescence obtained from normal sample): F2 (fluorescence obtained from tumor sample). In the case that the genomes of both samples are identical, then hybridization will be perfect, i.e. the number of base pairs present in both complementary strands are equivalent (usually denoted as a single line or one in the plot), however, if there is any problem in the sample (for example duplication of the sequences of any of the sample) there will be an increase in the intensity of the sample, as shown in figure 6.14.

6.5.2 Epigenetic profiling

Epigenetics involves changes in the regulation of the gene profile, in particular by means of biochemical changes (modifications or physical interaction) of DNA/ chromatin that result in heritable changes in gene expression and eventually affect gene regulation at the cell level [87]. These changes are not associated with changes in the DNA sequence, i.e. epigenetics is associated with changes in phenotype without changes in genotype. Eventually these changes influence the cells' innate approach to reading genes. Epigenetic profiling is the study of DNA methylation and histone modifications which result in changes in the accessibility of DNA to

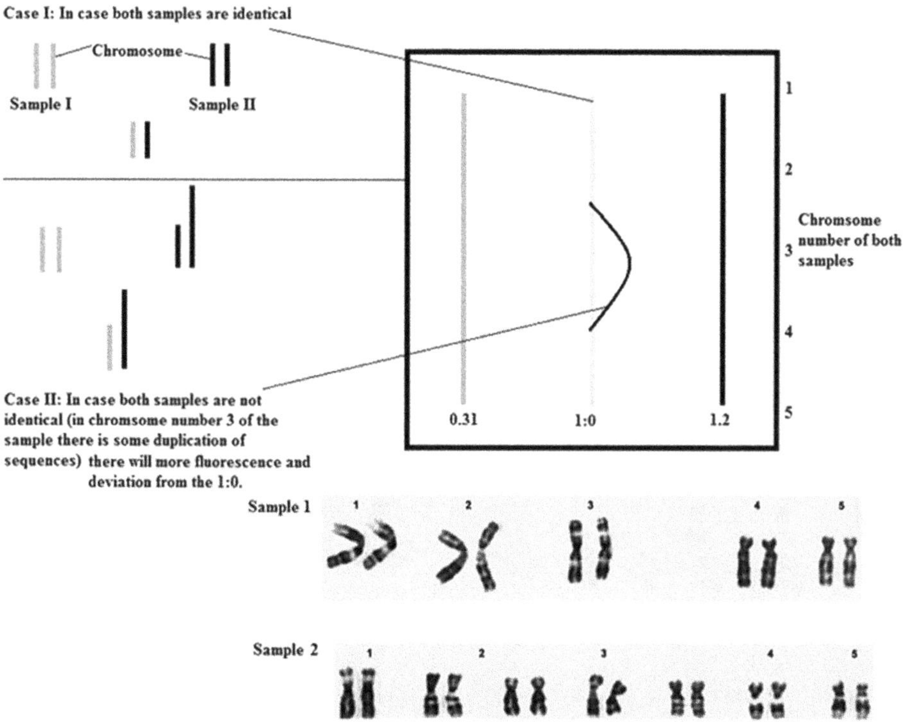

Figure 6.14. Comparative genomic hybridization.

the transcriptional machinery. Thus these heritable changes which regulate patterns of gene expression can be tracked by epigenetic profiling, e.g. the methylation of CpG dinucleotides and modifications of the tails of histone proteins. The procedure of epigenetic profiling involves certain key steps, as shown in figure 6.15. Step one involves histone protein modification, which is followed by binding of proteins at the modified histone sites. This modification is succeeded by crosslinking with formaldehyde. After crosslinking, the histone proteins, which are connected by chromatin fibers, are separated using sonification forces. When histone proteins are separated, the addition of antibody solution allows the binding of specific proteins (attached to modified histones) with specific antibodies. This process is called immunoprecipitation and allows protein purification. An antibody with the desirable protein is incubated to encourage antibody–protein binding in the solution, which will eventually result in the formation of an antibody–antigen complex. This complex can be removed by means of agarose beads, as mentioned in figures 6.15(a)–(d).

6.5.3 Fluorescence *in situ* hybridization (FISH)

FISH is a cytogenetic technique discovered in the 20th century. In this process fluorescent DNA probes are utilized to target specific chromosomal locations within the nucleus. This results in colored signals that can be examined using a

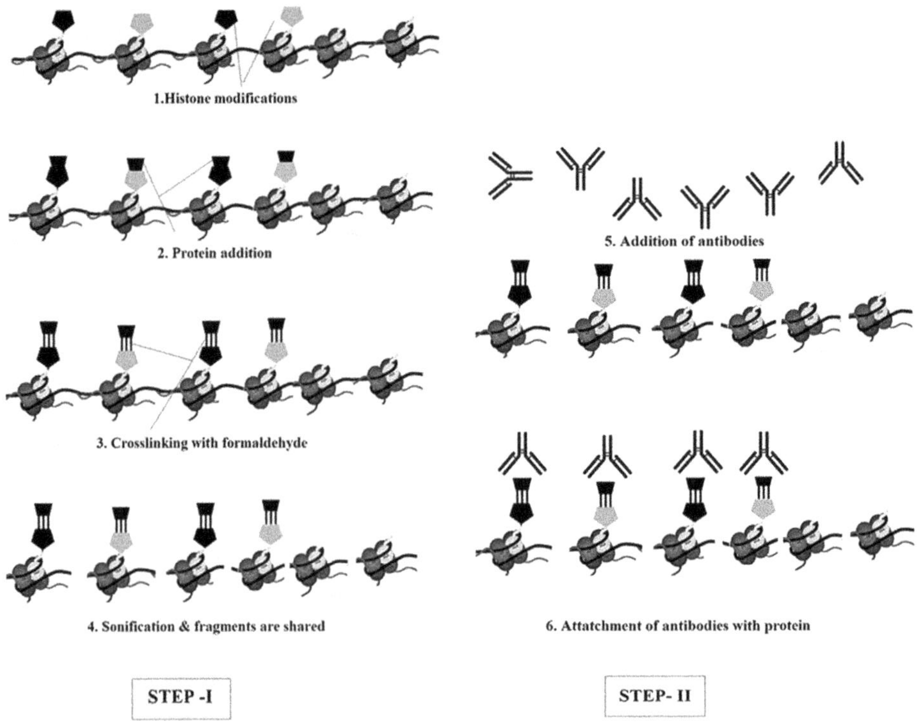

Figure 6.15. Epigenetic profiling: (a) step 1, (b) step 2, (c) step 3 and (d) step 4.

fluorescence microscope. In contrast to the earlier cytogenetic (CC) metaphase karyotype analysis (which relies on the presence of mitotically active dividing cells), FISH offers several advantages. For example, it does not require cell culturing, and can directly use fresh or paraffin-embedded interphase nuclei for rapid evaluation. Recent explorations of many disease-related genes have extended the applications of FISH to include more genetic diseases, hematologic malignancies and solid tumors. Currently FISH is being explored for targeted therapy in chronic myeloid leukemia, in particular for the detection of BCR/ABL1 translocation, HER2 amplification and ALK rearrangement [88–92]. Recent innovations also allow the use of FISH in the management and diagnosis of breast cancer [93, 94] and lung adenocarcinoma [93, 94]. Therefore, FISH examination is considered an important component of personalized medicine. With regard to biomarker discovery, several advanced high-throughput molecular tools, such as single nucleotide polymorphism arrays, array based comparative genome hybridization and next-generation sequencing, have been established and introduced into regular clinical practice. This development raises the question of the replacement of classical low-throughput assays, such as FISH, however, despite new advances, FISH is still considered an important diagnostic tool due to its simplicity and reliability in evaluating key biomarkers in various tumors [95]. The procedure of FISH is shown in figure 6.16.

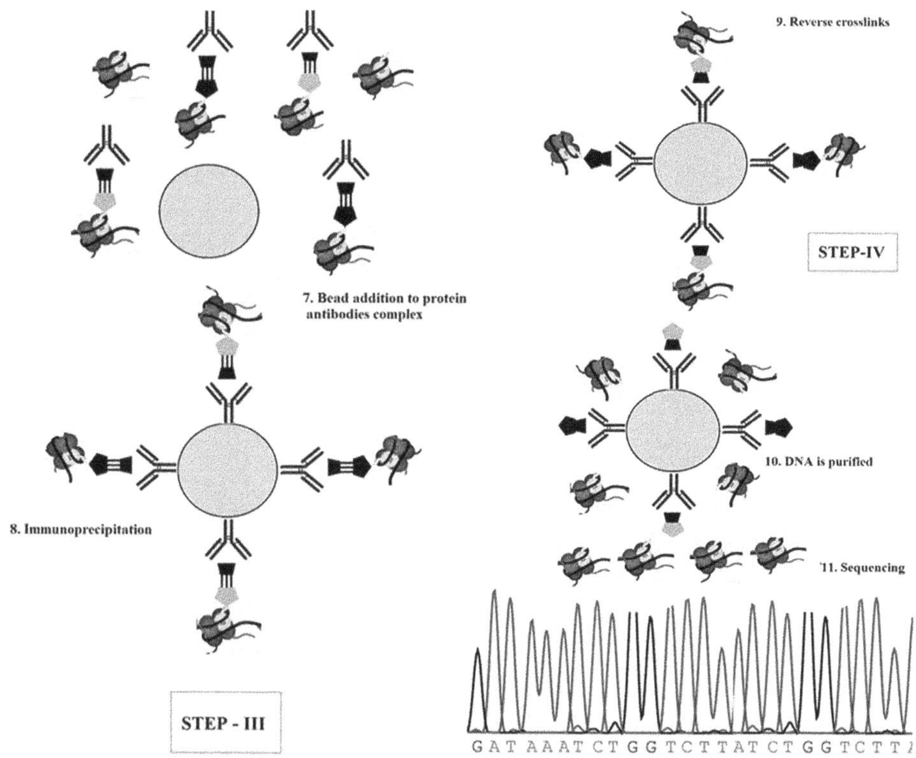

Figure 6.15. (Continued.)

6.5.4 Karyotyping

Karyotyping is the study of changes in chromosome number and their morphological variations, such as changes in size, the position of centromeres and banding patterns (light and dark bands obtained by staining the chromosome under a microscope). During sequential development, stem cells in culture can suffer from changes in chromosome structure and function, termed chromosomal abnormalities. Such types of changes can result in alterations of the genetic expression and cellular functions of stem cells [96]. This can eventually escalate the chances of the stem cells transforming from normal to tumorigenic. Therefore, it is important to routinely check cultures using karyotyping, mainly in stem cells intended for therapeutic use. The karyotyping procedure is shown in figure 6.17.

6.5.5 Single nucleotide polymorphism (SNP) analysis

Stem cells express both unique and specific combinations of transcription factors, cell surface proteins and cytoplasmic proteins. The techniques used for stem cell analysis and characterization include flow cytometry, array based analysis of the transcriptome, immunocytochemistry, western blots and biomarker analysis.

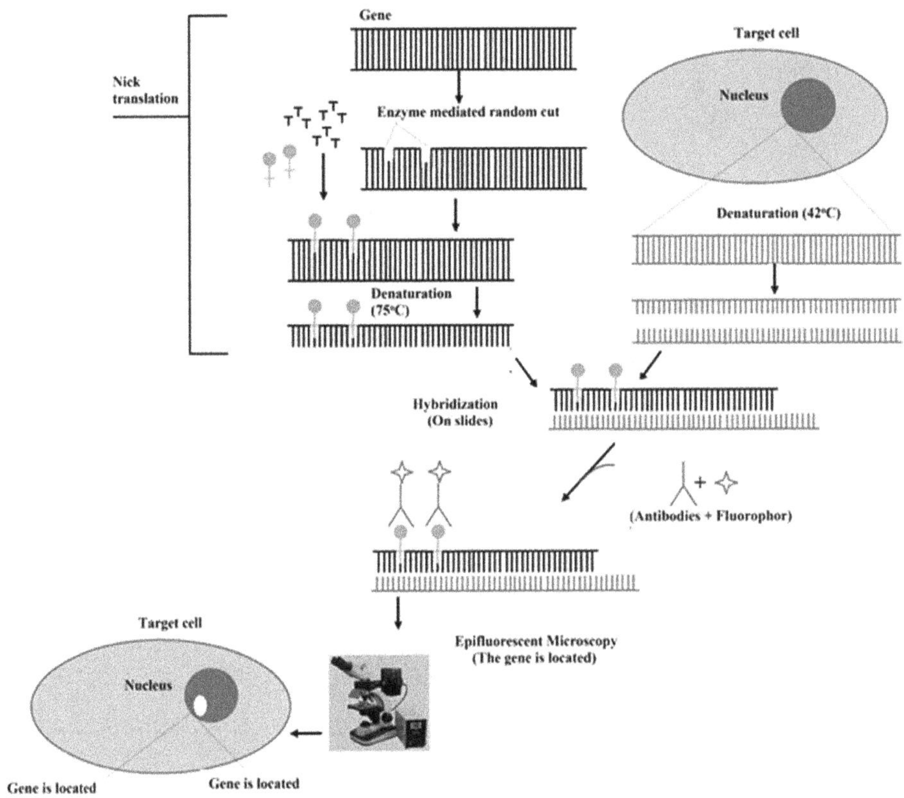

Figure 6.16. The FISH procedure.

6.5.6 Pluripotency markers (proteins)

Pluripotency is the cell's differentiation potential to develop into any cell type. Pluripotent stem cells, in particular embryonic stem cells, are derived from the inner cell mass (ICM) of the blastocyst (an early-stage embryo). Markers in this context are the molecules specifically expressed in embryonic stem cells. Knowledge related to the functions of these markers, i.e. the molecular mechanism involved, is important for elucidation and characterization of embryonic stem cells. However, different cell types can have single or occasionally multiple markers, therefore the key problem in the clinical application of embryonic stem cells is to purify embryonic stem cells from other types of cells, in particular tumor cells. Recently, advanced techniques such as the marker based flow cytometry technique and magnetic cell sorting have become the most effective cell isolation procedures. Pluripotency markers allow ease in identification as well as isolation of, in particular, tumor cells by means of these procedures. A wide range of cell surface and generic molecular markers are available that are indicative of the undifferentiated cells (figure 6.18). Pluripotency markers in the form of kits and reagents are available for the identification of pluripotent embryonic stem cells and induced pluripotent stem cells by immunocytochemical and enzyme activity analysis [97].

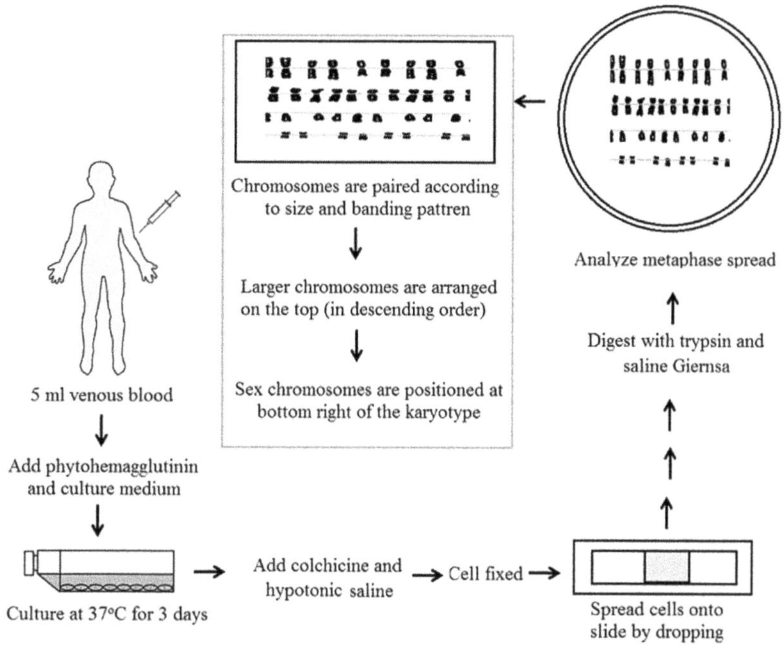

Figure 6.17. The karyotyping procedure.

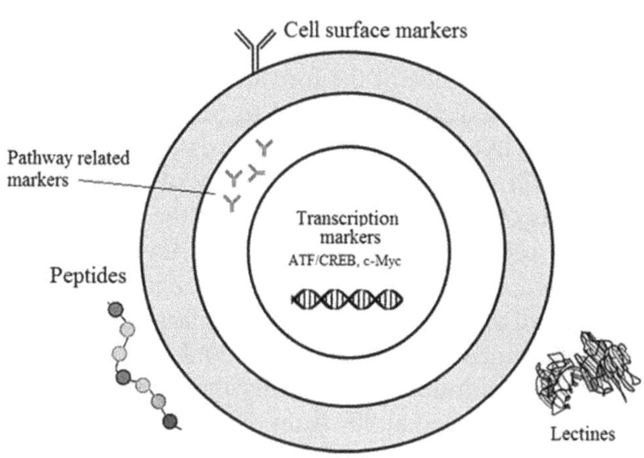

Figure 6.18. Pluripotency markers.

6.5.7 Stem cell arrays

A stem cell array contains a definite set of validated gene expression markers to characterize the identity of human embryonic stem cell and assess the phenotypic variations between embryonic stem cell isolates. Stem cell arrays contains miRNA, transferrin (TF), cDNA and gene expression.

6.5.8 Flow cytometry

Flow cytometry is a powerful technique for the analysis of multiple parameters of individual cells within a heterogeneous population. Flow cytometry has a wide range of applications from immunophenotyping to ploidy analysis, cell counting and GFP expression analysis [98–102]. A flow cytometer performs this analysis by passing thousands of cells per second through a laser beam and capturing the light that emerges from each cell as it passes through. Collected data can be analyzed statistically by flow cytometry software, which reports cellular characteristics such as size, complexity, phenotype and health [98–102]. The flow cytometer contains a fluidics system which presents the sample to the interrogation point and takes away the waste. Lasers operate as light sources for scattering and fluorescence, while optical components gather and collect the light for detectors which receive and record the light. The electronics and computer system then converts the signal from the detectors into digital data. The interrogation point is the heart of the system. This is where the sample and laser intersect and the optics collects the results and scattered fluorescence [98–102]. Each sample is transported to the interrogation point and for accurate data collection it is important that particles from the sample are passed through the laser beam one at a time. Most flow cytometers accomplish this by injecting the cell stream in a sheath fluid (the solution that runs in a flow cytometer) [98–102]. When the sheath fluid is running at laminar flow, the cells are injected into the center of the stream, at a slightly higher pressure. By doing this the sample is compressed to roughly one cell in diameter. This is called hydrodynamic focusing. As the cell passes through the laser it scatters or refracts light at various angles. Forward scatter and low angle light scatter is the amount of light scattered in forward direction. The magnitude of the forward scatter is roughly proportional to the size of the cell and this data can be used to quantify that parameter [98–102]. Scattered light received by the detectors is translated into a voltage pulse and, because smaller cells will produce smaller amounts of scatter and larger cells will produce larger amounts of scatter, the magnitude of the pulse recorded for each cell is proportional to the cell size. A histogram of the flow cytometry is a graphical representation of the size distribution within the population, but such a graph only presents one-dimensional data. Light scattering at a larger angle, which is also called side scattering, is determined by the granularity and structural complexity present inside the cell. This light is initially focused by the lens system and detected by a separate detector, usually located at 90° from the laser's path. The one-dimensional histogram does not necessarily show the complexity of the cell population, therefore a two-dimensional representation is more reliable in this context. Another parameter that can help in analyzing cell structure and function is fluorescence. Fluorescence is a term used to describe the excitation of a fluorophore to a higher energy level followed by the return of that fluorophore to the ground state through the emission of light. The energy of the emitted light is dependent on the energy at which the fluorophore was excited, and this light has a specific wavelength and consequently a specific color. One of the common approaches to studying cell characteristics using a flow cytometer is by using a fluorophore labeled antibody. In these experiments the fluorophore labeled antibody is added to the sample and the antibody then binds to a specific molecule present over the cell surface or inside the molecule. Finally, when laser light of a specific wavelength strikes the fluorophore, a fluorescence signal is emitted and detected by the detector of the flow cytometer. A typical flow cytometer with its basic features is shown in figure 6.19.

In flow cytometery microbes can be identified on the basis of:

- Their typical cytometric characteristics.
- Fluorochromes that can be used individually or with specific antibodies or oligonucleotides.

Flow cytometery has allowed the development of quantitative methods to examine the antimicrobial vulnerability and cytotoxicity of drugs in a highly reproducible, precise and rapid manner. In addition, this technique also allows the evaluation of *in vitro* antimicrobial activity and antimicrobial treatments *ex vivo*. One of the most important roles of flow cytometery is the assessment of heterogeneous populations with different responses against antimicrobial treatments. In clinical microbiology, flow cytometery can be used to detect organisms responsible for infections. Additionally, this procedure also allows the identification of micro-organisms, in

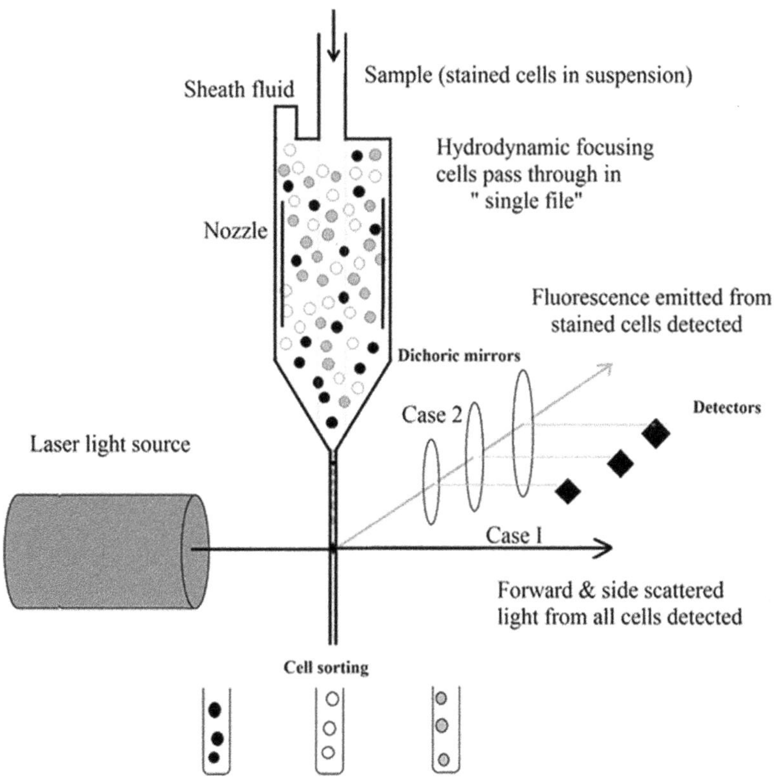

Figure 6.19. Flow cytometer.

particular on the basis of their cytometric characteristics. Flow cytometery allows testing for slow-growing micro-organisms, such as mycobacteria and fungi [102–111]. Applications of flow cytometery can also be used to study the immune response in patients, detect specific antibodies [102–111] and monitor clinical status after antimicrobial treatments [102–111]. In addition, when properly applied, flow cytometery can be used to define parameters that avoid subjectivity and assist the researcher in the analysis of specific results, mainly in the field of rapid diagnosis. The scattering of light in a flow cytometer is shown in figure 6.20.

6.6 Applications of animal cell cultures

Animal utilization is now restricted due to genuine moral and ethical concerns related to performing laboratory experiments on animals. Several animal welfare committees continue to criticize the utilization of animals in laboratories across the world. For these reasons, most researchers prefer to work on animal cell cultures for various studies. Some of the major applications of animal tissue cultures are described in table 6.11 and figure 6.21 [112–126].

6.6.1 Biopharmaceuticals from animal tissue culture

One of the most important applications of animal tissue culture is in the production of biopharmaceuticals for therapeutic use, as described in figure 6.21. Vaccine production was achieved using monkey kidney, chick embryo and human diploid

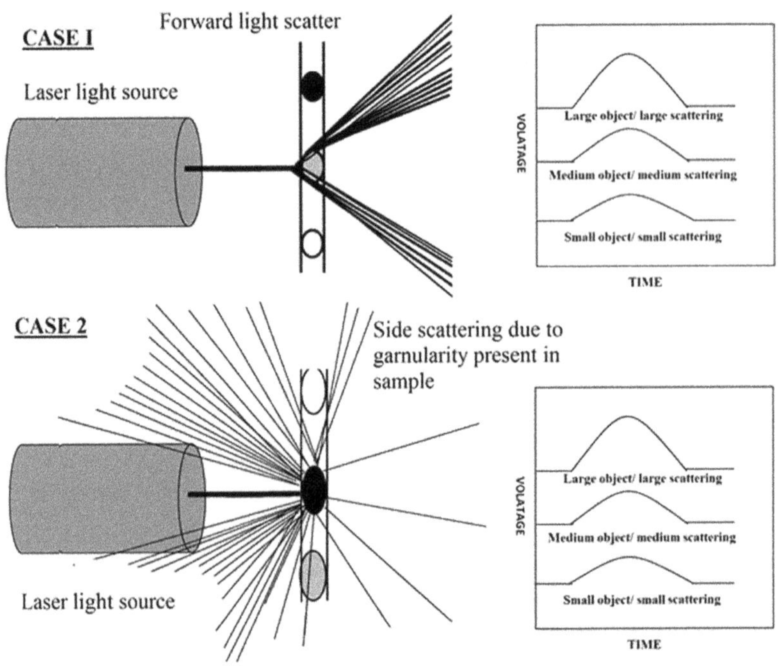

Figure 6.20. Scattering of light in a flow cytometer.

Table 6.11. Applications of animal tissue culture [112–126].

Category	Applications
Intracellular investigations	Investigations related to the cell cycle, differentiation, transcription, translation, and energy and drug metabolism.
Intracellular flux elucidation	Investigations related to hormonal receptors, metabolites, signal transduction and membrane trafficking.
Studying cell–cell interactions	Investigations related to cell adhesion and motility, matrix interaction, morphogenesis, paracrine control and metabolic cooperation.
Investigating environmental interactions	Investigations related to drug action, infection, cytotoxicity, mutagenesis and carcinogenesis.
Checking cellular programming	Biosynthetic pathways, expression of proteins, particular enzymes.
Genetic studies	Genetic analysis, transfection, transformation, immortalization and senescence.
Cell products	Production of cellular products such as monoclonal antibodies, vaccines, hormones and enzymes, as shown in figure 6.21.

cells. Vaccine production in animal tissue culture involves multiple procedures, which can increase the chances of contamination, and require attention to safety aspects. Due to this, most vaccines are produced by rDNA technology using bacteria and yeasts.

6.6.2 High value proteins

High value proteins, such as plasminogen activators, clotting factors and erythropoietins, can be produced using this technique. However, for their production post-translational modification, such as glycosylation, carboxylation, etc, is required, hence organisms with post-translational modification machinery are preferred. Organisms such as bacteria and yeast do not possess this machinery. However, biopharmaceuticals that do not require post-transitional modifications, such as insulin and growth hormone, can be produced by bacteria and yeast.

6.6.3 Co-culturing of mammalian cells and algae

Most algae contain bioactive secondary metabolites; these have a high commercial value. The many roles of these algae have been explored in different areas such as plant tissue culture science, medical science, parasitology and nanotechnology. The role of algae has rarely been studied in the context of animal tissue culture. Most green plants and algae synthesize adenosine triphosphate using various nutrients [127–158]. This helps reduce nicotinamide adenine dinucleotide phosphate, ultimately releasing O_2 using sunlight and water. Glyceraldehyde-3-phosphate is a

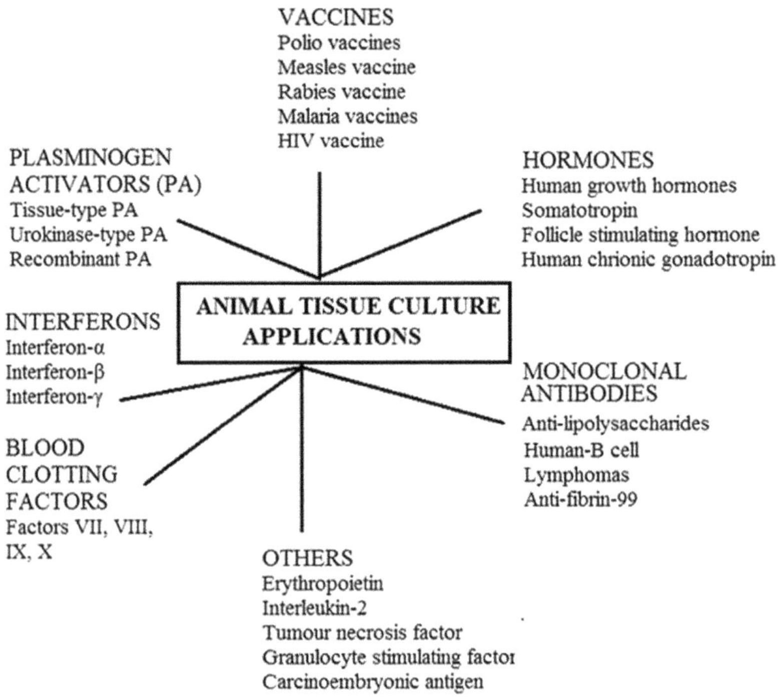

Figure 6.21. Applications of animal tissue culture in production of biopharmaceuticals.

compound which is produced during the carbon-fixation cycle from carbon dioxide and water by utilizing ATP and NADPH [127–158]. Most of the nutrients required for the algae's survival are synthesized by flyceraldehyde-3-phosphate. In contrast, animals, including humans, intake oxygen and all the required nutrients during biological respiration and produce ATP [127–158]. This biological respiration process releases carbon dioxide, which can be reutilized as a main component during photosynthesis. In agricultural practice, animal waste products are utilized as manure, which acts as an important source of nitrogen for crop plants. Similarly, glutamate, which is a metabolite of animals, is produced from 2-oxoglutarate and ammonia by green plants and algae. Glutamine, an amino acid, is produced from glutamate and ammonia. Animals also use these amino acids [127–158], thus a symbiotic relationship (co-culture) has to be established so that one organism meets the nutritional requirements of the other, in particular during *in vitro* culture. In fact, all organisms present on Earth can be considered to be in a symbiotic relationship, with the environmental 'recycling systems' of the various carbon–nitrogen–oxygen cycles. Biological tissues/organs derived from primary cells in animal tissue culture science can multiply under *in vitro* conditions and are frequently considered as potential sources of regenerative therapy. To repair injured tissues, several cell based regenerative and tissue-engineered treatments have been explored in the past few years [127–158]. One potential example is the development of a temperature-responsive culture surface, called cell sheet technology, which can be utilized as a culture surface to further develop cell–cell junctions, cell surface proteins, the extracellular matrix and cell-dense three-dimensional (3D) tissues [127–158]. These cellular sheets can be transplanted into numerous animal models with damaged tissues and allow the regeneration and restoration of their original physiological characteristics and functions. Several clinical investigations using single- or multi-layered cell sheets have been performed successfully [127–158]. Recent developments have allowed the design of 3D culture systems, as two-dimensional (2D) cultured cells are not able to sufficiently mimic the *in vivo* environment. The differences between the two environments disturb the gene expression and biochemical activity of the cells. Notably, a 3D culture system mimics the *in vivo* environment more closely than a 2D environment [127–158]. One of the major challenges of regenerative science is to closely mimic the *in vivo* environment by using functional 3D tissue, which can replace actual living tissues. Thus in tissue engineering, the construction of millimeter- or centimeter-sized three-dimensional (3D) human tissues with controlled 3D cell density, 3D cell components and 3D cell locations has been an aim of investigators. Although cell-dense 3D tissues can be fabricated by the simple layering of cell sheets, the ischemic (blood supply) environment makes the construction of thicker tissues difficult. Based on reports, the depth restriction of 3D tissues devoid of vascular networks is about 40–80 μm [127–158]. Severe hypoxia/under-nutrition within thicker multi-layered cell sheet tissues without vascular networks is likely, which can induce tissue damage [127–158]. The fabrication of thicker tissues is a long-standing objective of researchers in tissue engineering. Co-cultivation of mammalian cells and algae allows the construction of thicker tissues, which shows the potential of this process in the fields of cell biology, tissue engineering and regenerative medicine. This symbiotic *in vitro* system between mammalian cells and algae allows the development of a recycling system in which the algae supply oxygen to mammalian cells, and reuse metabolites and waste products from mammalian cells, while mammalian cells consume the oxygen, and excrete metabolites and waste products.

References

[1] Zegers M M P, O'Brien L E, Yu W, Datta A and Mostov K E 2003 Epithelial polarity and tubulogenesis *in vitro* *Trends Cell Biol.* **13** 169–76

[2] Wells R G 2008 The role of matrix stiffness in regulating cell behavior *Hepatology* **47** 1394–400

[3] Giaever I and Keese C R 1993 A morphological biosensor for mammalian cells *Nature* **366** 591–2

[4] Giaever I and Keese C R 1991 Micromotion of mammalian cells measured electrically *Proc. Natl Acad. Sci. USA* **81** 3761–4

[5] Giaever I and Keese C R 1986 Use of electric fields to monitor the dynamical aspect of cell behavior in tissue culture *IEEE Proc. Biomed. Eng.* **33** 242–7

[6] Kim S A, Tai C Y, Mok L P, Mosser E A and Schuman E M 2011 Calcium-dependent dynamics of cadherin interactions at cell–cell junctions *Proc. Natl Acad. Sci. USA* **108** 9857–62

[7] Hirano S, Nose A, Hatta K, Kawakami A and Takeichi M 1987 Calcium-dependent cell–cell adhesion molecules (cadherins): subclass specificities and possible involvement of actin bundles *J. Cell Biol.* **105** 2501–10

[8] Kamiya K, Tsumoto K, Yoshimura T and Akiyoshi K 2011 Cadherin-integrated liposomes with potential application in a drug delivery system *Biomaterials* **32** 9899–907

[9] Brackenbury R, Rutishauser U and Edelman G M 1981 Distinct calcium-independent and calcium-dependent adhesion systems of chicken embryo cells *Proc. Natl Acad. Sci. USA* **78** 387–91

[10] Wight T N, Kinsella M G and Qwarnström E E 1992 The role of proteoglycans in cell adhesion, migration and proliferation *Curr. Opin. Cell Biol.* **4** 793–801

[11] Stacey G N, Byrne E and Hawkins J R 2014 DNA profiling and characterization of animal cell lines *Methods Mol. Biol.* **1104** 57–73

[12] O'Connor C 2008 Fluorescence *in situ* hybridization (FISH) *Nature Educ.* **1** 171

[13] Janeway C A Jr *et al* 2001 The major histocompatibility complex and its functions *Immunobiology: The Immune System in Health and Disease* 5th edn (New York: Garland Science)

[14] Young H Y, Zucker P, Flavell R A, Jevnikar A M and Singh B 2004 Characterization of the role of major histocompatibility complex in type 1 diabetes recurrence after islet transplantation *Transplantation* **78** 509–15

[15] Rugg-Gunn P J *et al* 2012 Cell-surface proteomics identifies lineage-specific markers of embryo-derived stem cells *Dev. Cell* **22** 887–901

[16] Draper J S, Pigott C, Thomson J A and Andrews P W 2002 Surface antigens of human embryonic stem cells: changes upon differentiation in culture *J. Anat.* **200** 249–58

[17] Barak V 1, Goike H, Panaretakis K W and Einarsson R 2004 Clinical utility of cytokeratins as tumor markers *Clin. Biochem.* **37** 529–40

[18] Pinkus G S and Kurtin P J 1985 Epithelial membrane antigen—a diagnostic discriminant in surgical pathology: immunohistochemical profile in epithelial, mesenchymal, and hematopoietic neoplasms using paraffin sections and monoclonal antibodies *Hum. Pathol.* **16** 929–40

[19] Krone B, Lenz A, Heermann K H, Seifer M, Lu X Y and Gerlich W H 1990 Interaction between hepatitis B surface proteins and monomeric human serum albumin *Hepatology* **11** 1050–6

[20] Svensson M, Sabharwal H, Håkansson A, Mossberg A K, Lipniunas P, Leffler H, Svanborg C and Linse S 1999 Molecular characterization of alpha-lactalbumin folding variants that induce apoptosis in tumor cells *J. Biol. Chem.* **274** 6388–96

[21] Hatakeyama K, Wakabayashi-Nakao K, Ohshima K, Sakura N, Yamaguchi K and Mochizuki T 2013 Novel protein isoforms of carcinoembryonic antigen are secreted from pancreatic, gastric and colorectal cancer cells *BMC Res. Notes* **6** 381

[22] Lilja H, Ulmert D and Vickers A J 2008 Prostate-specific antigen and prostate cancer: prediction, detection and monitoring *Nat. Rev. Cancer* **8** 268–78

[23] van de Stolpe A and van der Saag P T 1996 Intercellular adhesion molecule-1 *J. Mol. Med.* **74** 13–33

[24] Adigun O O and Bhimji S S 2018 *Alpha Fetoprotein (AFP, Maternal Serum Alpha Fetoprotein, MSAFP)* (Treasure Island, FL: StatPearls)

[25] Bahl O P 1977 Human chorionic gonadotropin, its receptor and mechanism of action *Fed. Proc.* **36** 2119–27

[26] Postel-Vinay M C and Finidori J 1995 Growth hormone receptor: structure and signal transduction *Eur. J. Endocrinol.* **133** 654–9

[27] Ivaska J 2011 Vimentin: central hub in EMT induction? *Small GTPases* **2** 51–3

[28] Akiyama S K 1996 Integrins in cell adhesion and signaling *Hum. Cell* **9** 181–6

[29] Dominguez R A 2000–2013 Common binding site for actin-binding proteins on the actin surface *Madame Curie Bioscience Database* (Austin, TX: Landes Bioscience)

[30] Dunehoo A L, Anderson M, Majumdar S, Kobayashi N, Berkland C and Siahaan T J 2006 Cell adhesion molecules for targeted drug delivery *J. Pharm. Sci.* **95** 1856–72

[31] Halliday G M, Cullen K M, Kril J J, Harding A J and Harasty J 1996 Glial fibrillary acidic protein (GFAP) immunohistochemistry in human cortex: a quantitative study using different antisera *Neurosci. Lett.* **209** 29–32

[32] Eng L F 1985 Glial fibrillary acidic protein (GFAP): the major protein of glial intermediate filaments in differentiated astrocytes *J. Neuroimmunol.* **8** 203–14

[33] Mayer C A, Brunkhorst R, Niessner M, Pfeilschifter W, Steinmetz H and Foerch C 2013 Blood levels of glial fibrillary acidic protein (GFAP) in patients with neurological diseases *PLoS One* **8** e62101

[34] Liang S L and Chan D W 2007 Enzymes and related proteins as cancer biomarkers: a proteomic approach *Clin. Chim. Acta* **381** 93–7

[35] Mythili S and Malathi N 2015 Diagnostic markers of acute myocardial infarction *Biomed. Rep.* **3** 743–8

[36] Wu A H 1999 Cardiac markers: from enzymes to proteins, diagnosis to prognosis, laboratory to bedside *Ann. Clin. Lab. Sci.* **29** 18–23

[37] Drexler H G and Gaedicke G 1985 Significance of enzyme markers as a part of multiple marker analysis in leukemia research *Am. J. Pediatr. Hematol. Oncol.* **7** 283–9

[38] Howell K E, Ito A and Palade G E 1978 Endoplasmic reticulum marker enzymes in Golgi fractions—what does this mean? *J. Cell Biol.* **79** 581–9

[39] Mitchell L 2011 A tale of two repressors *J. Mol. Biol.* **409** 14–27

[40] Rasmuson M, Rasmuson B and Nilson L R 1967 A study of isoenzyme polymorphism in experimental populations of *Drosophila melanogaster Hereditas* **57** 263–74

[41] Byrne D H 1990 Isozyme variability in four diploid stone fruits compared with other woody perennial plants *J. Hered.* **81** 68–71

[42] Torres A M 1990 Isozyme analysis of tree fruits *Isozymes in Plant Biology* ed D E Soltis and P S Soltis (London: Chapman and Hall) pp 192–205

[43] Arulsekar S, Parfitt D E and Kester D E 1986 Comparison of isozyme variability in peach and almond cultivars *J. Hered.* **77** 272–4

[44] Goldberg R F *et al* 2008 Intestinal alkaline phosphatase is a gut mucosal defense factor maintained by enteral nutrition *Proc. Natl Acad. Sci. USA* **105** 3551–6

[45] Caldwell P R, Seegal B C, Hsu K C, Das M and Soffer R L 1976 Angiotensin-converting enzyme: vascular endothelial localization *Science* **191** 1050–1

[46] Baird M F, Graham S M, Baker J S and Bickerstaff G F 2012 Creatine-kinase- and exercise-related muscle damage implications for muscle performance and recovery *J. Nutr. Metab.* **2012** 960363

[47] Bürklen T S *et al* 2006 The creatine kinase/creatine connection to Alzheimer's disease: CK inactivation, APP-CK complexes, and focal creatine deposits *J. Biomed. Biotechnol.* **2006** 35936

[48] Martinez-Hernandez A, Bell K P and Norenberg M D 1977 Glutamine synthetase: glial localization in brain *Science* **195** 1356–8
[49] Strober W 2001 Wright-Giemsa and nonspecific esterase staining of cells *Curr. Protoc. Immunol.* **11** A.3D.1-A.3D.4
[50] Traber P G, Yu L, Wu G D and Judge T A 1992 Sucrase–isomaltase gene expression along crypt-villus axis of human small intestine is regulated at level of mRNA abundance *Am. J. Physiol.* **262** G123–30
[51] Clarkson K, Sturdgess I and Molyneux A 2001 The usefulness of tyrosinase in the immunohistochemical assessment of melanocytic lesions: a comparison of the novel T311 antibody (anti-tyrosinase) with S-100, HMB45, and A103 (anti-melan-A) *J. Clin. Pathol.* **54** 196–200
[52] Rusciano D, Lorenzoni P, Lin S and Burger M M 1998 Hepatocyte growth factor/scatter factor and hepatocytes are potent downregulators of tyrosinase expression in B16 melanoma cells *J. Cell Biochem.* **71** 264–76
[53] Tahara E and Yokozaki H 1988 Growth factors as biological markers of malignancy *Gan To Kagaku Ryoho* **15** 1102–8
[54] Stacker S A, Williams R A and Achen M G 2004 Lymphangiogenic growth factors as markers of tumor metastasis *APMIS* **112** 539–49
[55] Bianchi E *et al* 2013 Growth factors, their receptor expression and markers for proliferation of endothelial and neoplastic cells in human osteosarcoma *Int. J. Immunopathol. Pharmacol.* **26** 621–32
[56] Nicolaides K H 2013 Screening for chromosomal defects *Ultrasound Obst. Gyn.* **21** 313–21
[57] Ujvari B and Belov K 2011 Major histocompatibility complex (MHC) markers in conservation biology *Int. J. Mol. Sci.* **12** 5168–86
[58] Bacchetti S and Graham F L 1977 Transfer of the gene for thymidine kinase to thymidine kinase-deficient human cells by purified herpes simplex viral DNA *Proc. Natl Acad. Sci. USA* **74** 1590–4
[59] Gilbert D A *et al* 1990 Application of DNA fingerprints for cell-line individualization *Am. J. Hum. Genet.* **47** 499–514
[60] Callaghan R, Luk F and Bebawy M 2014 Inhibition of the multidrug resistance P-glycoprotein: time for a change of strategy? *Drug Metab. Dispos.* **42** 623–31
[61] Marshall P N 1978 Reticulation, polychromasia and stippling of erythrocytes *Microsc. Acta* **81** 89–106
[62] Jager M M, Murk J L, Piqué R D, Hekker T A and Vandenbroucke-Grauls C M 2010 Five-minute Giemsa stain for rapid detection of malaria parasites in blood smears *Trop. Doct.* **41** 33–5
[63] Houwen B 2002 Blood film preparation and staining procedures *Clin. Lab. Med.* **22** 1–14
[64] Comings D E 1978 Mechanisms of chromosome banding and implications for chromosome structure *Annu. Rev. Genet.* **12** 25–46
[65] Francke U 1992 Review: *Chromosome Banding* by Sumner A T (London: Unwin Hyman) (1990) *Cell* **68** P1005–6
[66] Bickmore W A 2001 *Karyotype Analysis and Chromosome Banding* (Chichester: eLS Wiley)
[67] Daga R R, Thode G and Amores A 1996 Chromosome complement, C-banding, Ag-NOR and replication banding in the zebrafish *Danio rerio Chromosome Res.* **4** 29–32
[68] Schreck R R and Distèche C M 2001 Chromosome banding techniques *Curr. Protoc. Hum. Genet.* **4** 4.2.1–4.2.36

[69] Ried T, Schröck E, Ning Y and Wienberg J 1998 Chromosome painting: a useful art *Hum. Mol. Genet.* **7** 1619–26
[70] Xu S S, Liu Z, Zhang Q, Niu Z, Jan C C and Cai X 2016 Chromosome painting by GISH and multicolor FISH *Methods Mol. Biol.* **1429** 7–21
[71] Masabanda J S and Griffin D K 2003 Generation of chromosome paints: approach for increasing specificity and intensity of signals *Biotechniques* **34** 530–2 534, 536
[72] Carter N P 1994 Cytogenetic analysis by chromosome painting *Cytometry* **18** 2–10
[73] Belloni E, Lahortiga I, Odero M, Di Fiore P and Pelicci P 2010 Spectral karyotyping (SKY) *Ecancer. Med. Sci.* **4** 181
[74] Imataka G and Arisaka O 2011 Chromosome analysis using spectral karyotyping (SKY) *Cell Biochem. Biophys.* **62** 13–7
[75] Belaud-Rotureau M A, Elghezal H, Bernardin C, Sanlaville D, Radford-Weiss I, Raoul O, Vekemans M and Romana S P 2003 Spectral karyotyping (SKY) principle, advantages and limitations *Ann. Biol. Clin.* **61** 139–46
[76] Liyanage M *et al* 1996 Multicolour spectral karyotyping of mouse chromosomes *Nat. Genet.* **14** 312–5
[77] Schröck E and Padilla-Nash H 2000 Spectral karyotyping and multicolor fluorescence *in situ* hybridization reveal new tumor-specific chromosomal aberrations *Semin. Hematol.* **37** 334–47
[78] Roewer L 2013 DNA fingerprinting in forensics: past, present, future *Invest. Genet.* **4** 22
[79] Daiger S P 1991 DNA fingerprinting *Am. J. Hum. Genet.* **49** 899–903
[80] Signer E, Kuenzle C C, Thomann P E and Hübscher U 1988 DNA fingerprinting: improved DNA extraction from small blood samples *Nucl. Acids Res.* **16** 7738
[81] Matos L L, Trufelli D C, de Matos M G, da Silva and Pinhal M A 2010 Immunohistochemistry as an important tool in biomarkers detection and clinical practice *Biomark Insights* **5** 9–20
[82] Okuno Y, Sasao F, Fukunaga T and Fukai K 1977 An application of PAP (peroxidase-anti-peroxidase) staining technique for the rapid titration of dengue virus type 4 infectivity *Biken J.* **20** 29–33
[83] Weiss M M *et al* 1999 Comparative genomic hybridisation *Mol. Pathol.* **52** 243–51
[84] Houldsworth J and Chaganti R S 1994 Comparative genomic hybridization: an overview *Am. J. Pathol.* **145** 1253–60
[85] Redon R and Carter N P 2009 Comparative genomic hybridization: microarray design and data interpretation *Methods Mol. Biol.* **529** 37–49
[86] Pinkel D and Albertson D G 2005 Comparative genomic hybridization *Annu. Rev. Genomics Hum. Genet.* **6** 331–54
[87] Martín-Subero J I and Esteller M 2011 Profiling epigenetic alterations in disease *Adv. Exp. Med. Biol.* **711** 162–77
[88] Wertheim G B, Hexner E and Bagg A 2012 Molecular-based classification of acute myeloid leukemia and its role in directing rational therapy: personalized medicine for profoundly promiscuous proliferations *Mol. Diagn. Ther.* **2** 357–69
[89] Landau D A and Wu C J 2013 Chronic lymphocytic leukemia: molecular heterogeneity revealed by high-throughput genomics *Genome Med.* **2** 47
[90] Dietel M *et al* 2013 Predictive molecular pathology and its role in targeted cancer therapy: a review focussing on clinical relevance *Cancer Gene Ther.* **2** 211–21

[91] Boyd C and Boyle D P 2012 Molecular diagnosis on tissues and cells: how it affects training and will affect practice in the future *Cytopathology* **2** 286–94
[92] Jiang H *et al* 2012 The utility of fluorescence *in situ* hybridization analysis in diagnosing myelodysplastic syndromes is limited to cases with karyotype failure *Leuk. Res.* **2** 448–52
[93] Pathmanathan N and Bilous A M 2012 HER2 testing in breast cancer: an overview of current techniques and recent developments *Pathology* **2** 587–95
[94] Kim H R *et al* 2013 The frequency and impact of ROS1 rearrangement on clinical outcomes in never smokers with lung adenocarcinoma *Ann. Oncol.* **2** 2364–70
[95] Casaluce F, Sgambato A, Maione P, Rossi A, Ferrara C, Napolitano A, Palazzolo G, Ciardiello F and Gridelli C 2013 ALK inhibitors: a new targeted therapy in the treatment of advanced NSCLC *Target Oncol.* **2** 55–67
[96] Gandhi L and Janne P A 2012 Crizotinib for ALK-rearranged non-small cell lung cancer: a new targeted therapy for a new target *Clin. Cancer Res.* **2** 3737–42
[97] Zhao W, Ji X, Zhang F, Li L and Ma L 2012 Embryonic stem cell markers *Molecules* **17** 6196–236
[98] Brown M and Wittwer C 2000 Flow cytometry: principles and clinical applications in hematology *Clin. Chem.* **46** 1221–9
[99] Han Y, Gu Y, Zhang A C and Lo Y H 2016 Review: imaging technologies for flow cytometry *Lab Chip* **16** 4639–47
[100] Alvarez D F, Helm K, Degregori J, Roederer M and Majka S 2009 Publishing flow cytometry data *Am. J. Physiol. Lung Cell Mol. Physiol.* **298** L127–30
[101] Dawn M 2015 Betters use of flow cytometry in clinical practice *J. Adv. Pract. Oncol.* **6** 435–40
[102] Mund J A, Estes M L, Yoder M C, Ingram D A and Case J 2012 Flow cytometric identification and functional characterization of immature and mature circulating endothelial cells *Arterioscler. Thromb. Vasc. Biol.* **32** 1045–53
[103] Gant V A, Warnes G, Phillips I and Savidge G F 1993 The application of flow cytometry to the study of bacterial responses to antibiotics *J. Med. Microbiol.* **39** 147–54
[104] Kirk S M, Callister S M, Lim L C and Schell R F 1997 Rapid susceptibility testing of *Candida albicans* by flow cytometry *J. Clin. Microbiol.* **35** 358–63
[105] Pore R S 1994 Antibiotic susceptibility testing by flow cytometry *J. Antimicrob. Chemother.* **34** 613–27
[106] Walberg M, Gaustad P and Steen H B 1997 Rapid discrimination of bacterial species with different ampicillin susceptibility levels by means of flow cytometry *Cytometry* **29** 267–72
[107] Patterson B K, Mosiman V L, Cantarero L, Furtado M, Bhattacharya M and Goolsby C 1998 Detection of HIV-RNA-positive monocytes in peripheral blood of HIV-positive patients by simultaneous flow cytometric analysis of intracellular HIV RNA and cellular immunophenotype *Cytometry* **31** 265–74
[108] Cohen C Y and Sahar E 1989 Rapid flow cytometric bacterial detection and determination of susceptibility to amikacin in body fluids and exudates *J. Clin. Microbiol.* **27** 1250–6
[109] Hu Y W, Birch P, Balaskas E, Zeibdawi A, Scalia V, Thériault-Valin S A, Gill P and Aye M T 1996 Flow cytometric immunofluorescence assay for detection of antibodies to human immunodeficiency virus type 1 using insoluble precursor forms of recombinant polyproteins as carriers and antigens *J. Clin. Microbiol.* **34** 1412–9
[110] Alvarez-Barrientos A, Arroyo J, Cantón R, Nombela C and Sánchez-Pérez M 2000 Applications of flow cytometry to clinical microbiology *Clin. Microbiol. Rev.* **13** 167–95

[111] Best L M, Veldhuyzen van Zanten S J, Bezanson G S, Haldane D J and Malatjalian D A 1992 Serological detection of *Helicobacter pylori* by a flow microsphere immunofluorescence assay *J. Clin. Microbiol.* **30** 2311–7
[112] Khan K H 2013 Gene expression in mammalian cells and its applications *Adv. Pharm. Bull.* **3** 257–63
[113] Butler M 2005 Animal cell cultures: recent achievements and perspectives in the production of biopharmaceuticals *Appl. Microbiol. Biotechnol.* **68** 283–91
[114] Crameri G *et al* 2009 Establishment, immortalisation and characterization of pteropid bat cell lines *PLoS One* **4** e8266
[115] Prouty S M, Hanson K D, Boyle A L, Brown J R, Shichiri M, Follansbee M R, Kang W and Sedivy J M 1993 A cell culture model system for genetic analyses of the cell cycle by targeted homologous recombination *Oncogene* **8** 899–907
[116] Hudu S A, Alshrari A S, Syahida A and Sekawi Z 2016 Cell culture, technology: enhancing the culture of diagnosing human diseases *J. Clin. Diagn. Res.* **10** DE01–5
[117] Alberts B *et al* 2002 Cell–cell adhesion *Molecular Biology of the Cell* 4th edn (New York: Garland Science)
[118] Khalili A A and Ahmad M R 2015 A review of cell adhesion studies for biomedical and biological applications *Int. J. Mol. Sci.* **16** 18149–84
[119] Fuchs M *et al* 2003 Adhesion, cell motility and the cytoskeleton *Signal Transduct.* **3** 158–63
[120] Sánchez-Kopper A, Becker M, Pfizenmaier J, Kessler C, Karau A and Takors R 2016 Tracking dipeptides at work-uptake and intracellular fate in CHO culture *AMB Express.* **6** 48
[121] Young J D 2013 Metabolic flux rewiring in mammalian cell cultures *Curr. Opin. Biotechnol.* **24** 1108–15
[122] Sheikholeslami Z, Jolicoeur M and Henry O 2014 Elucidating the effects of postinduction glutamine feeding on the growth and productivity of CHO cells *Biotechnol. Prog.* **30** 535–46
[123] Sims J K, Manteiga S and Lee K 2013 Towards high resolution analysis of metabolic flux in cells and tissues *Curr. Opin. Biotechnol.* **24** 933–9
[124] Niklas J, Schräder E, Sandig V, Noll T and Heinzle E 2011 Quantitative characterization of metabolism and metabolic shifts during growth of the new human cell line AGE1.HN using time resolved metabolic flux analysis *Bioprocess Biosyst. Eng.* **34** 533–45
[125] Amaral A I, Teixeira A P, Håkonsen B I, Sonnewald U and Alves P M 2011 A comprehensive metabolic profile of cultured astrocytes using isotopic transient metabolic flux analysis and 13C-labeled glucose *Front. Neuroenergetics* **3** 5
[126] Kocan R M, Sabo K M and Landolt M L 1985 Cytotoxicity/genotoxicity: the application of cell culture techniques to the measurement of marine sediment pollution *Aquat. Toxicol.* **6** 165–77
[127] Haraguchi Y, Kagawa Y, Sakaguchi K, Matsuura K, Shimizu T and Okano T 2017 Thicker three-dimensional tissue from a 'symbiotic recycling system' combining mammalian cells and algae *Sci. Rep.* **7** 41594
[128] Albert B *et al* 2015 Energy conversion: mitochondria and chloroplasts *Molecular Biology of the Cell* vol 6 ed B Albert *et al* (New York: Garland Science) ch 14, pp 753–812
[129] Haraguchi Y, Shimizu T, Yamato M and Okano T 2012 Cell therapy and tissue engineering for cardiovascular disease *Stem. Cells Transl. Med.* **1** 136–41
[130] Behfar A, Crespo-Diaz R, Terzic A and Gersh B J 2014 Cell therapy for cardiac repair-lessons from clinical trials *Nat. Rev. Cardiol.* **11** 232–46

[131] Okano T, Yamada N, Sakai H and Sakurai Y 1993 A novel recovery system for cultured cells using plasma-treated polystyrene dishes grafted with poly (N-isopropylacrylamide) *J. Biomed. Mater. Res.* **27** 1243–51

[132] Haraguchi Y *et al* 2012 Fabrication of functional three-dimensional tissues by stacking cell sheets *in vitro Nat. Protoc.* **7** 850–8

[133] Haraguchi Y *et al* 2014 Cell sheet technology for cardiac tissue engineering *Methods Mol. Biol.* **1181** 139–55

[134] Memon I A *et al* 2005 Repair of impaired myocardium by means of implantation of engineered autologous myoblast sheets *J. Thorac. Cardiovasc. Surg.* **130** 1333–41

[135] Sekine H *et al* 2011 Cardiac cell sheet transplantation improves damaged heart function via superior cell survival in comparison with dissociated cell injection *Tissue Eng.* A **17** 2973–80

[136] Chang D *et al* 2015 Time course of cell sheet adhesion to porcine heart tissue after transplantation *PLoS One* **10** e0137494

[137] Nishida K *et al* 2004 Corneal reconstruction with tissue-engineered cell sheets composed of autologous oral mucosal epithelium *N. Engl. J. Med.* **351** 1187–96

[138] Haraguchi Y, Shimizu T, Yamato M and Okano T 2011 Regenerative therapies using cell sheet-based tissue engineering for cardiac disease *Cardiol. Res. Pract.* **2011** 845170

[139] Sawa Y *et al* 2012 Tissue engineered myoblast sheets improved cardiac function sufficiently to discontinue LVAS in a patient with DCM: report of a case *Surg. Today* **42** 181–4

[140] Burillon C *et al* 2012 Cultured autologous oral mucosal epithelial cell sheet (CAOMECS) transplantation for the treatment of corneal limbal epithelial stem cell deficiency *Invest. Ophthalmol. Vis. Sci.* **53** 1325–31

[141] Ohki T *et al* 2012 Prevention of esophageal stricture after endoscopic submucosal dissection using tissue-engineered cell sheets *Gastroenterology* **143** 582–8

[142] Sato M, Yamato M, Hamahashi K, Okano T and Mochida J 2014 Articular cartilage regeneration using cell sheet technology *Anat. Rec.* **297** 36–43

[143] Egami M, Haraguchi Y, Shimizu T, Yamato M and Okano T 2014 Latest status of the clinical and industrial applications of cell sheet engineering and regenerative medicine *Arch. Pharm. Res.* **37** 96–106

[144] Iwata T *et al* 2015 Cell sheet engineering and its application for periodontal regeneration *J. Tissue Eng. Regen. Med.* **9** 343–56

[145] Yaguchi Y *et al* 2016 Middle ear mucosal regeneration with three-dimensionally tissue-engineered autologous middle ear cell sheets in rabbit model *J. Tissue Eng. Regen.* **10** E188–94

[146] Abbott A 2004 Biology's new dimension *Nature* **424** 870–2

[147] Kaneshiro N *et al* 2006 Bioengineered chondrocyte sheets may be potentially useful for the treatment of partial thickness defects of articular cartilage *Biochem. Biophys. Res. Commun.* **349** 723–31

[148] Mitani G *et al* 2009 The properties of bioengineered chondrocyte sheets for cartilage regeneration *BMC Biotechnol.* **9** 17

[149] Shimizu T *et al* 2006 Polysurgery of cell sheet grafts overcomes diffusion limits to produce thick, vascularized myocardial tissues *FASEB J.* **20** 708–10

[150] Sekine W, Haraguchi Y, Shimizu T, Umezawa A and Okano T 2011 Thickness limitation and cell viability of multi-layered cell sheets and overcoming the diffusion limit by a porous-membrane culture insert *J. Biochip. Tissue chip.* **S2** 001

[151] Haraguchi Y *et al* 2010 Development of a new assay system for evaluating the permeability of various substances through three-dimensional tissue *Tissue Eng.* C **16** 685–92
[152] Schneider M, Marison I W and von Stockar U 1996 The importance of ammonia in mammalian cell culture *J. Biotechnol.* **46** 161–85
[153] Bhatia S 2015 *Modern Applications of Plant Biotechnology in Pharmaceutical Sciences* (London: Academic, Elsevier)
[154] Bhatia S 2016 *Natural Polymer Drug Delivery Systems* (Basel: Springer)
[155] Bhatia S 2016 *Systems for Drug Delivery* (Basel: Springer)
[156] Bhatia S 2016 *Nanotechnology in Drug Delivery Fundamentals, Design, and Applications* (Boca Raton, FL: CRC Press)
[157] Bhatia S 2018 *Introduction to Pharmaceutical Biotechnology: Enzymes, Proteins and Bioinformatics* vol 2 (Bristol: IOP Publishing)
[158] Bhatia S 2016 *Introduction to Pharmaceutical Biotechnology: Basic Techniques and Concepts* vol 1 (Bristol: IOP Publishing)

IOP Publishing

Introduction to Pharmaceutical Biotechnology, Volume 3
Animal tissue culture and biopharmaceuticals
Saurabh Bhatia, Tanveer Naved and Satish Sardana

Chapter 7

Role of pharmacists in delivering biotechnological products

7.1 Introduction

Biotechnology is a broad science, with numerous applications in different disciplines derived from the different concepts and techniques employed, such as using living systems for the production of biopharmaceuticals or other products, e.g. yeast for the production of alcoholic beverages and bacteria to produce pharmaceuticals [1, 2]. Currently, various biopharmaceutical proteins are also manufactured in bacteria and yeast [3, 4]. Recent progress in biotechnological science led to the development of recombinant DNA processes, or hybridoma technology, which yields various new products every year. These biotherapeutics are widely used for the diagnosis, treatment and monitoring of the many ailments. Moreover, there is potential for these biotherapeutics to be utilized in a number of medical subspecialties, including endocrinology, cardiology, oncology, immunology and infectious diseases. Products which are obtained from biotechnological processes and have significance in the pharmaceutical sector are called biopharmaceuticals. Since these products are different from the usual pharmaceutical products, a lack of information regarding their prescription, dispensing and compounding hinder their utilization by health practitioners and pharmacists. During a general examination/survey for pharmaceutical products, sometimes it is very difficult for the patient/pharmacist/physician to obtain detailed information based on the trade name, however, if such products can be examined based on their basic salt or standard active pharmaceutical ingredient (API), it will be easy to obtain detailed information about their efficacy, safety, storage, mechanism of action and other relevant information. However, when dealing with biopharmaceuticals (whose active ingredients are proteins) their complex mechanisms, varying adverse drug reactions and degradation issues, in addition to the lack of information on administration and procurement procedures, have slowed their wider promotion and adoption. Owing to these

doi:10.1088/2053-2563/aafac0ch7

factors pharmacists and other health professionals should be adequately trained to handle such products effectively. The biopharmaceuticals sector generates immense opportunities for pharmacists to care for patients undergoing therapy, in particular in an ambulatory care setting. Additionally, this allows the pharmacist to become more involved in patient outcomes. Pharmacists also have a duty to promote awareness regarding all aspects of biopharmaceuticals, which may promote the role of pharmacist towards new horizons of pharmaceutical care. For example, when dealing with radiopharmaceuticals the pharmacist should actively participate in promoting awareness of the safe use of these agents. However, often the pharmacist is replaced with a non-pharmacy worker, which may further reduce the opportunities in this sector. Animal biotechnology is a new area that is growing at a great pace. The biopharmaceuticals developed from this field require an active role for the pharmacist, with their knowledge and awareness of these products. Recently, various animal biotechnology based products have been explored, among them some belonging to biopharmaceuticals. In comparison to general pharmaceutical products, only a few biopharmaceuticals are present on the market, and there is no specific professional course available to train individuals with an interest in biopharmaceuticals. Therefore, special education and training programs are required to assist future pharmacists to easily deal with different types of biopharmaceuticals. In this chapter we discuss some of the active roles of pharmacists dealing with biotechnological products.

7.2 The role of the pharmacist

During the course of biopharmaceutical therapy, the pharmacist should have enough understanding of the patient's needs to deliver the therapeutic agent in the minimum possible time period. To achieve this, a collaborative effort between pharmacists, manufacturers/wholesalers/retailers, physicians and insurance agents is required. Such collaboration allows the delivery of the product in the minimum time span. Additionally, the pharmacist can also take the guidance of physicians if the patient shows intolerance or unusual side effects after the first biopharmaceutical therapy. A change of therapy can be achieved under the instruction of a physician, health adviser or practitioner who has past experience dealing with biopharmaceuticals. Some of the basic relationships of the pharmacist during this therapy process are illustrated in figure 7.1.

7.3 The route of administration for biopharmaceuticals

There are different types of delivery systems that are suitable for the delivery of biopharmaceuticals (figures 7.2 and 7.3). According to reports, most biopharmaceuticals are delivered through injections, usually through the intravenous, subcutaneous or intramuscular route. Administration through the parenteral route is sometimes promoted in this context, as it overcomes some issues [5–7]. However, such administration requires a special training program and it is very difficult for the patient to self-administer. Additionally administration through intramuscular injection may cause several types of injuries, e.g. abscesses, hematoma and severe allergic

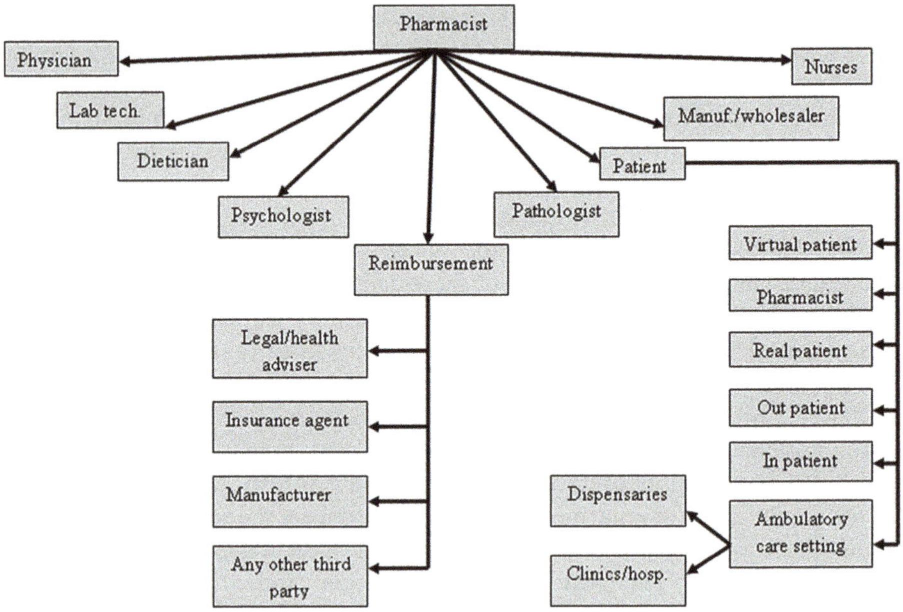

Figure 7.1. The relationships between the pharmacist and different professionals during biopharmaceutical therapy.

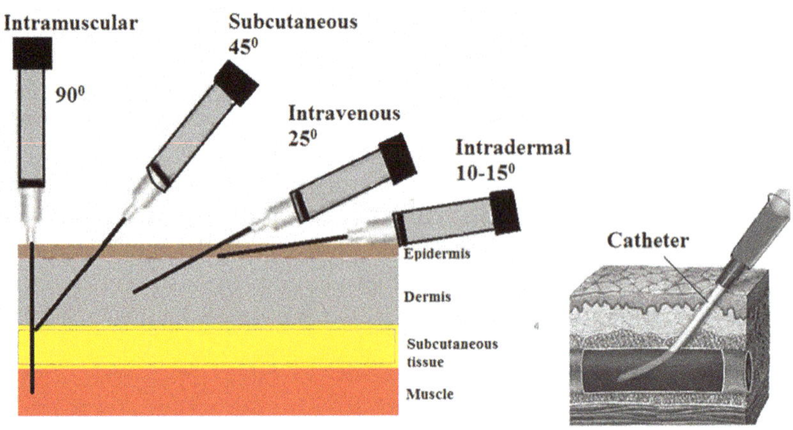

Figure 7.2. Angles of insertion of parenteral injections.

reactions (figure 7.2). Administration through injections can be very difficult, since one needs to take into account any physicochemical limitations, gastric and enzymatic degradation, and the potential for reduced immune reactions, while ensuring molecular stability and permeability [5–7].

The associated cost and patient discomfort place limits on applications and promote alternative possibilities that increase patient compliance. Current advances have led to various alternative routes being proposed for the delivery of

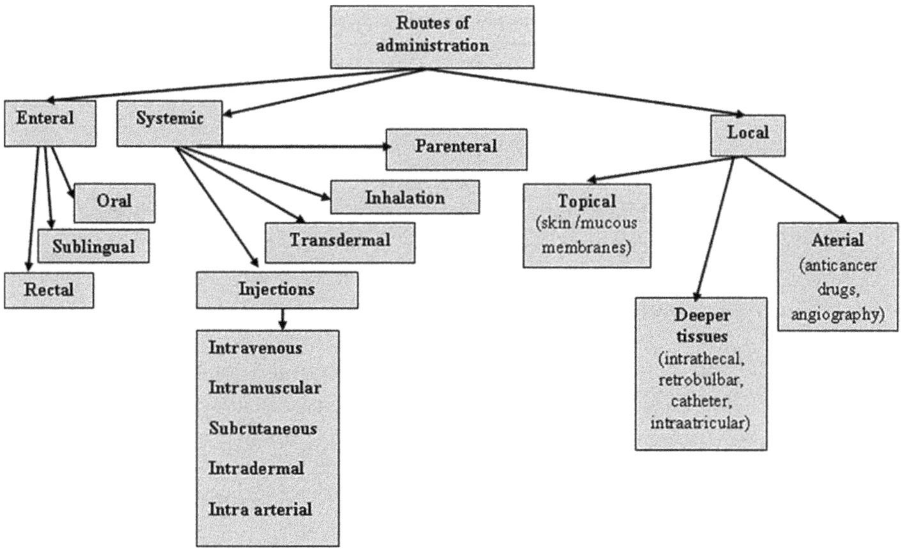

Figure 7.3. Various routes of administration of biopharmaceuticals.

Table 7.1. Biopharmaceuticals with alternative routes of administration [5–7].

Technology/manufacturer	Product and delivery mode
Biphasix™ (Helix BioPharma)	Transdermal delivery using liposomes.
ImuXen® (Lipoxen)	Oral delivery of DNA and vaccines.
AERx® (Aradigm Corporation)	Inhaled insulin by pulmonary delivery (aerosol delivery system).
Technosphere® (Mankind Corporation)	Insulin-loaded dry powder microspheres (phase III clinical trials). Delivered by inhalation.
Exubera® (Pfizer)	Dry powder formulation of insulin (withdrawn from market). Exubera is a spray dried insulin powder in unit dose blisters and a pulmonary inhaler, a medical device.

biopharmaceutical drugs, such as the oral, transdermal, pulmonary, nasal, vaginal, sublingual, rectal and ocular delivery routes (table 7.1). These routes are all accepted for systemic biopharmaceutical delivery. However, the poor acceptability of some of the routes, such as the vaginal one, limits their utilization to local use only [5–7]. Thus the oral, nasal, pulmonary and transdermal routes are still acknowledged as the leading alternative drug delivery routes, with some technologies already available commercially (table 7.1). Different biopharmaceuticals and their respective modes of delivery are listed in table 7.2.

Table 7.2. Modes of delivery and the respective biopharmaceutical.

Mode of delivery	Procedure	Biopharmaceutical	Ref.
Epidural	Administration of injection or infusion into the epidural space.	TNF inhibitor, CEP 37247, in a phase II program	[10]
Intracerebral	Administration into the cerebrum, direct injection into the brain.	CpG oligodeoxynucleotide	[11]
	Intracerebroventricular	Administration into cerebral ventricles/ administration into the ventricular system of the brain.	
	Octadecaneuropeptide	[12]	
Epicutaneous or topical	Administration into the skin.	Viaskin®, protein antigen TNP-Ig	[13]
Sublingual and buccal administration	Placement under the tongue.	VIAtab™ oral formulation of insulin, Desmopressin (approved product)	[6]
Extra-amniotic administration	Administration between the endometrium and fetal membranes	Prostaglandin F2	[14]
Nasal administration	Administration through the nose can be used for topically acting substances, as well as for insufflation.	Influenza virus vaccine (FluMist), salmon calcitonin, (approved product)	[6]
Intra-arterial	Administration into an artery.	Albumin microspheres for intra-arterial tumor targeting	[15]
Intra-articular	Administration into a joint space.	Gene transfer	[17]
Intracardiac	Administration into the heart.	Stem cell injection	[18]
Intracavernous injection	An injection into the base of the penis.	Invicorp®, Aviptadil, i.e. a vasoactive intestinal polypeptide	
Intradermal	Administration into the skin.	Influenza virus vaccine	
Intralesional	Administration into a skin lesion, for treatment of local skin lesions.	Interleukin-2	[19, 20]

(*Continued*)

Intramuscular	Administration into a muscle.	Hepatitis B immune globulin, rabies immune globulin, respiratory syncytial virus, monoclonal antibody, recombinant hepatitis A vaccine	
Intraocular	Administration into the eye.	Aganirsen GS-101 (an antisense oligonucleotide or short naturally occurring DNA strand), siRNA, antisense oligonucleotides, ribozymes, aptamers	
Intraosseous infusion	Administration into the bone marrow.	Insulin infusion	[22]
Intraperitoneal	Infusion or injection into the peritoneum.	Interferoninterferon α-2b	[23–25]
Intravitreal	Administration through the eye.	Erytheropoietin, an antisense nucleotide polymer against CMV retinitis in patient with AIDS	[26, 27]
Subcutaneous	Administration under the skin.	Insulin, interferon β-1a (Avonex (Biogen) and Rebif (Merck)), INTRON® A (interferon α-2b), blood clotting factor, CSF, antibodies, interleukins, enzymes, hormones, vaccines, interferons	[28]
Intrathecal/ intraspinal	Administration into the spinal canal.	Blood clotting factor, CSF, antibodies, interleukins, enzymes, hormones, vaccines, interferons	[29]
Intravaginal administration	Administration in the vagina.	Interferon, antigen-carrying vaccine nanoparticles	[30, 31]
Intravenous	Administration into a vein.	Blood clotting factor, CSF, antibodies, interleukins, enzymes, hormones, vaccines, interferons	[32]

Intravesical	Infusion into the urinary bladder.	rhGM-CSF	[33]
Transdermal	Diffusion through the intact skin for systemic rather than topical distribution.	PTH, insulin (in clinical trials), interferon-gamma (IFN-gamma) mediated by penetratin, a cell-permeable peptide	[6, 34]
Intrathecal	Injection into the spinal canal.	A humanized IgG4 antibody, bapineuzumab, and solane-zuma	[35]
	Intracerebroventricular	Administration into the cerebral ventricular system.	
Interleukin-1	[36]		
Intracerebral	Administration into the brain parenchyma.	Interferon-gamma	[37]
Transmucosal	Diffusion through a mucous membrane.	Recombinant human interferon-α B/D hybrid	[38]
Inhalational	Via respiratory route/lungs.	Insulin, dornase α (approved)	[6]
Intranasal	Via nasal route.	Calcitonin for Pagels disease	[32]
Intrarespiratory	Via respiratory tract.	DNAse administered to lungs to reduce mucus accumulation	[32]
Topical	Local effect (applied to a localized area of the body or to the surface of a body part).	Platelet growth factor for wound healing	[32]

7.4 Development of biopharmaceuticals

There are three major factors that influence the development of biopharmaceuticals: manufacturing, safety, and the pharmacological and biological activity of the biopharmaceuticals (figure 7.4) [8]. During manufacturing, some factors such as the productivity, quality and formulabilty of biopharmaceuticals are important, since these factors play an important role in maintaining the stability of biopharmaceuticals (figure 7.4) [8]. Second, safety issues such as immunogenicity, hypersensitivity, immune-toxicological reactions and specificity play an important role in determining the safety profile of biopharmaceuticals [8]. Similarly, the evaluation of factors governing pharmacological and biological activities, such as delivery/ROA, mode of action and therapeutic efficacy, decide the final therapeutic potential and associated adverse drug reactions of biopharmaceuticals.

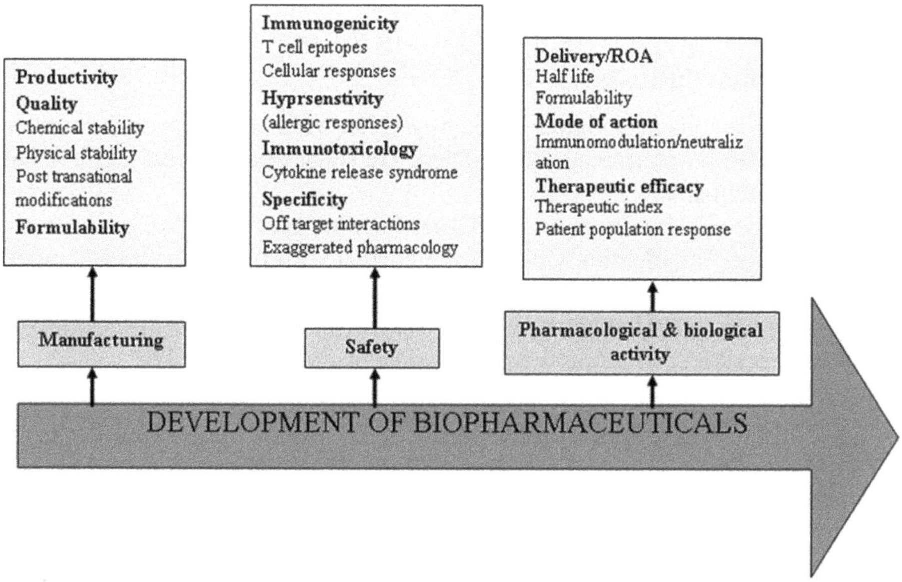

Figure 7.4. Development of biopharmaceuticals [8].

7.5 Basic requirements for biotechnology based products

Biotechnology based products are difficult to handle in a conventional manufacturing environment. Large molecular weight proteins can be sensitive to environmental and processing conditions (temperature, mixing time and speed, order of addition of formulation components, pH adjustment and control, and contact time with various surfaces such as filters and tubing). The quality evaluation of the compatibility between a desired biopharmaceutical and its production process occurs during actual experimentation, in particular through batch preparations that lead to process validation batch studies. It is essential to carefully evaluate the biopharmaceutical product activity, stability and overall quality to ensure the feasibility of scale-up studies for desired batches. Stability studies should be conducted to determine the factors that can affect the product life, which ultimately ensure the procurement conditions. All biotherapeutics and prepared formulations should be tested for compatibility with the production environment conditions, processing steps and equipment. The general procedures that are required to prepare sterile products constitute a series of events beginning with the procurement of approved raw materials and primary packaging components, and ending with a sterile product sealed in its dispensing package. The protocol for the unit processes involved in manufacturing finished sterile dosage forms include compounding and mixing, filtration, filling, lyophilization (freeze-drying), closing and sealing, sorting and inspection, labeling, and final packaging for distribution. During the formulation and development of biopharmaceuticals each step must be monitored and validated carefully to ensure product quality, particularly for relatively unstable and interactive biopharmaceuticals. Various biopharmaceuticals have been approved (251 approved products) [9] and some are in the clinical pipeline.

7.6 Strategies adopted by pharmacists for dealing with biopharmaceuticals

Pharmacists can consult a database or any other valid information source when dealing with with biopharmaceuticals. There are some good sources of information on biopharmaceuticals:

- Biotechnology Information and Issues for Pharmacists, Office of Professional Programs, Philadelphia College of Pharmacy and Science, Woodland Ave, Philadelphia, PA.
- Biotechnology Medicines in Development, Communications Division, Pharmaceutical Manufacturers Association, Washington, DC.
- Certain web links: biosimilarspipeline.com, www.biopharma.com, www.bioplanassociates.com.
- Current Biotechnology Abstracts (computerized database).
- De Muth J An Introduction to Pharmaceutical Biotechnology. Extension Services in Pharmacy, School of Pharmacy, University of Wisconsin-Madison, 1990.
- FDA Biopharmaceutical Approvals and US Biopharmacopeia Registry of Biopharmaceutical Products.
- Pharma Intelligence 1998 The pink sheet: prescription pharmaceuticals and biotechnology *FDC Reports,* Chevy Chase, MD
- Walsh G and Murph B 2013 *Biopharmaceuticals, an Industrial Perspective* (Berlin: Springer).
- Herbert W J, Wilkinson P C and Scott D L (ed) 1985 *Dictionary of Immunology* 3rd edn (Boston, MA: Blackwell Scientific)
- Pezzuto J M, Johnson M E and Manasse H R 2013 *Biotechnology and Pharmacy* (Berlin: Springer)
- Geigert J 2012 *The Challenge of CMC Regulatory Compliance for Biopharmaceuticals* (Berlin: Springer)
- Grindley J and Ogden J 1999 *Understanding Biopharmaceuticals: Manufacturing and Regulatory Issues* (London: Taylor and Francis)
- Pharmaceutical Biotechnology Monitor, Global Medical Communications, New York, NY.
- Pharmacy Practice News, pharmacypracticenews.com.
- Ho R J Y and Gibaldi M 2004 *Biotechnology and Biopharmaceuticals: Transforming Proteins and Genes into Drugs* (New York: Wiley)
- Pharma Intelligence, SCRIP World Pharmaceutical News, https://scrip.pharmaintelligence.informa.com/
- Pharma Intelligence, The NDA Pipeline *FDC Reports* Chevy Chase, MD; a searchable database that contains the latest information on over 900 pharmaceutical, biotechnological and related companies and their more than 7000 approval records.
- The Tufts Center (http://csdd.tufts.edu/research/databases) maintains a unique databases that provide the most detailed source of historical information available on pharmaceutical and biopharmaceutical innovation in the United States. These databases contain data on products approved from 1963 to the present.

Pharmacists and other related professionals can refer to these informative sources and guide patients, as well as other health professionals, accordingly. These informative sources should be present where these biopharmaceuticals are stocked or handled. By using such sources the pharmacist can update herself/himself regarding recent developments in biopharmaceuticals. It has been thirty years since the approval of the first biopharmaceutical (humulin, approved in 1982) [39]. Current biotechnology industries have matured rapidly in the intervening years. A recent publication reported the development of two hundred twenty products for general medical use and several hundred more in the pipeline [40]. With such developments, the global biopharmaceutical market was expected to increase to 167 billion US dollars by 2015 [41]. Currently, biopharmaceuticals include peptides, proteins, glycoproteins and nucleic acid based biological products. These biotherapeutics can potentially be used for therapeutic, prophylactic or *in vivo* diagnostic purposes [42]. The products described below are ideal for treatments outside the hospital environment, however, they are not all used in ambulatory care settings.

7.6.1 Utilization of hematopoietic growth factors

According to previous findings, blood is one of the most extremely regenerative tissues, with approximately one trillion (10^{12}) cells arising daily in adult human bone marrow [29]. Hematopoietic growth factors are the family of hormones that can regulate both the hematopoietic and the functional activity of mature blood cells [42]. A list of recombinant hematopoietic growth factors and their clinical uses in several applications are provided in table 7.3.

As the name suggests, colony-stimulating factors are therapeutic agents that promote the proliferation of blood cells. In standard practice they are used as an adjunct to chemotherapy. Usually they are administered subcutaneously to cancer patients in a similar way as a diabetic patient administers insulin. The patient must be trained in the proper technique of subcutaneous administration of colony-stimulating factors [45–48]. The most important cytokines include colony-stimulating factors, such as granulocyte macrophage colony-stimulating factor (GM-CSF), granulocyte

Table 7.3. Recombinant hematopoietic growth factors and their clinical uses in several applications [42].

HGFs s	Commercial name	Approved application
EPO (thrombopoietin)	Eprex® (Epoetin-α) Epogen® (Epoetin-β)	Anemia, reduction of allogenic blood transfusion in surgery patients.
G-CSF	Neupogen® (Filgrastim) Neulasta® (Pegfilgrastim)	Acute myeloid leukemia, severe chronic neutropenia.
GM-CSF	Leukine® (Sargramostim)	Acute myelogenous leukemia; myeloid recovery after autologous and allogenic bone marrow transplantation.
	Macrogen® (Molgramostim)	Severe neutropenia.

colony-stimulating factor (G-CSF), macrophage colony-stimulating factor (MCSF) and multi-CSF (or IL-3). Granulocyte macrophage colony-stimulating factor (GM-CSF) is a haematopoietic growth factor with a wide variety of applications in the clinic. In early phase I studies the continuous intravenous (c.i.) route of administration was often used. Later it was shown that subcutaneous administration was also effective and may be the preferred route of administration. However, in another report it was clearly indicated that intravenous granulocyte colony-stimulating factor (G-CSF) might be safer and more convenient than subcutaneous administration for hospitalized hemato-oncological patients receiving chemotherapy [45–48]. The application of granulocyte colony-stimulating factor in intensification chemotherapy for improvement of the prognosis for children with acute lymphoblastic leukemia has also been reported, but results in considerable morbidity, primarily due to myelosuppression with resultant neutropenia limiting its usage. Administration of recombinant granulocyte colony-stimulating factor (G-CSF) was found to shorten neutropenia following intensive chemotherapy [45–48].

7.6.2 Training for intramuscular administration

Those who are dependent on biopharmaceuticals require proper training and instruction for intramuscular administration of the correct dose. In the case of children, the parents are trained so that they will be able to administer the agent to their child. Intramuscular injection is a technique used to deliver medication deep into the muscles. This allows the medication to be absorbed into the bloodstream quickly [5]. You may have received an intramuscular injection at a doctor's office the last time you got a vaccine, such as the flu shot. Intramuscular injections are used to deliver drugs and vaccines. They are common practice in modern medicine. Several drugs and almost all inactivated vaccines are delivered this way. Intramuscular injections are used when other types of delivery methods are not recommended [5]. These include oral (swallowed into the stomach), intravenous (injected into the vein) and subcutaneous (injected just under the layer of skin). The speed of absorption is faster for intramuscular injection compared to subcutaneous injection. This is because the muscle tissue has a greater blood supply than the area just under the skin. Muscle tissue may also hold a larger volume of medication than subcutaneous tissue. Intramuscular injection may be used instead of intravenous injection because some drugs are irritating to veins. Sometimes, a suitable vein cannot be located. It may be used instead of oral delivery because some drugs are destroyed by the digestive system when a drug is swallowed. Intramuscular injections can be administered through the deltoid muscle of the arm, the vastus lateralis muscle of the thigh, and the entrogluteal and dorsogluteal muscles of the buttocks. The uptake of drugs from the thigh is slower than from the arm [5]. The thigh may be used when the other sites are not available or if a patient needs to administer the medication on their own while lying down. A ventrogluteal site such as the buttocks is the safest site for adults and children over seven months of age. It is deep and not close to any major blood vessels and nerves. The thigh is preferred for immunizations in children. The deltoid in the arm is the site most typically used for vaccines. It is rarely used for other medications because the volume of medication is too large. All healthcare workers who administer intramuscular injections to patients should receive training and education on proper administration. The selection of needle size and injection site depends on many factors. These include the age and size of the patient, and the volume of medication. An injection site should not be used if there is evidence of infection or injury at the site. The medication is administered with a needle long enough to reach the muscle without penetrating the structures underneath. The following general procedure may be used for intramuscular injections [5]:

- The practitioner positions the patient in a way that is comfortable and keeps the muscles relaxed.
- The site selected for injection is cleaned with an alcohol swab. The skin is allowed to air dry.
- The practitioner spreads out the skin of the selected site to isolate the muscle.
- The needle is inserted into the muscle at about a 90 degree angle.
- The syringe is often drawn back slightly to ensure that the needle has not penetrated a blood vessel.
- The vaccine is injected slowly into the tissue.
- The needle is withdrawn quickly and properly discarded.
- Light pressure is applied to the injection site using a dry cotton ball or gauze.
- The patient is monitored for a short time for swelling, bleeding or allergic reactions.

Complications of intramuscular injections include [5]:

- Abscess.
- Allergic reaction.
- Bleeding.
- Hematoma.
- Infection.
- Injury to blood vessels and peripheral nerves.
- Many children often experience anxiety related to the use of needles and fear of pain. A nurse practitioner should use extra precautions around children to help ease the child's worries and prevent injury.
- Pain at the injection site.
- Tingling or numbness.

Protein based macromolecules are solely administered by parenteral routes, such as intravenous, subcutaneous or intramuscular injection. There are several factors such as molecular size, hydrophilicity and gastric degradation that prevents gastrointestinal absorption of therapeutic proteins [5]. The subcutaneous route of administration of therapeutic proteins is known to be the most convenient and frequently preferred route. More specifically, the subcutaneous route of self-administration translates into considerably lower treatment costs. Therapeutic protein absorption from subcutaneous administration is likely to be slower when compared to small molecules as the absorption rates depend on the size of the molecule. For example, selecting subcutaneous injection for peptide administration in humans gives a maximum systemic concentration (T_{max}) in hours, whereas this increases to several days in the case of monoclonal antibodies. Generally factors such as body weight, sex, age, activity level and the influence of subcutaneous absorption parameters need to be taken into account. Moreover, factors such as skin morphology and physiology (the presence or absence of the panniculus carnosus muscle in the skin) also affect the rate of absorption of drugs through the subcutaneous route. Therefore, subcutaneous injection is not ideal for all species [5]. Some other factors such as subcutaneous blood flow, drug substance and product characteristics (the presence of Fc receptors (see below), target interactions, charge, formulation, dose concentration, total dose), and mode of administration (injection site, injection time, depth of injection, anesthesia status) also affect subcutaneous administration [5].

7.6.3 Pegnology and biopharmaceuticals

Pegnology is defined as the conjugation or linking of a therapeutic protein with polyethylene glycol (PEG) [49]. Various delivery systems have been explored for the delivery of therapeutic proteins at particular sites in a controlled or sustained manner. However, several side effects or shortcomings have restricted their use for their intended purposes. Drawbacks such as drug leakage, RES uptake, stability, immunogenicity, hemolytic toxicity, etc, can generally be overcome by PEGylation [49]. One of the main advantages of PEGylated therapeutic proteins is that they can be developed in several forms, such as proteins, liposomes, enzymes, drugs, nano-particles, etc. In this process PEG is covalently or non-covalently conjugated with biological proteins or drugs, or vesicles, which are then described as PEGylated (pegylated) [49]. This process is usually carried out by incubation of a reactive derivative of PEG with the sample compound or molecule. Conjugation, coating or attachment of PEG with the sample by formation of a covalent bond can potentially mask the agent from the host's immune system [49]. This masking can not only reduced immunogenicity and antigenicity, but also increases the hydrodynamic size (size in solution) of the agent, which prolongs its circulatory time by reducing renal clearance. PEGylation offers water solubility to hydrophobic drugs and proteins. This process can be applied over an array of biopharmaceuticals, recombinant proteins and other biologically active proteins, most typically peptides, proteins and antibody fragments, that can improve the safety and efficiency of many therapeutics. This process causes significant alterations in physiochemical properties, such as changes in hydrophobicity, conformation, electrostatic binding, etc. These changes in physiochemical properties enhance the systemic retention of the therapeutic agent. Additionally, it can affect the binding potential of the therapeutic moiety present on the targeted sample, which can ultimately affect the targeted bimolecular cell receptor interaction, and may further alter the absorption and distribution patterns. PEGylation enhances the molecular weight of a sample and potentially imparts several significant biological benefits over the unmodified form, such as:

- Enhanced protection from proteolytic degradation.
- Extended circulating life.
- Improved drug solubility.

- Increased drug stability.
- Reduced dosage frequency, without diminished efficacy and with potentially reduced toxicity.

Commercial benefits of PEGylated drugs:
- Opportunities for new delivery formats and dosing regimens.
- Extended patent life of previously approved drugs.

PEG is a particularly an attractive polymer for conjugation and due to its specific characteristics that are relevant to pharmaceutical applications:
- Altered distribution in the body.
- High mobility in solution.
- Lack of toxicity and low immunogenicity.
- Ready clearance from the body.
- Water solubility.

Currently, pegnology is widely employed to coat different therapeutic proteins. With such an advanced process, it is now not only possible to coat therapeutic proteins, but also allow an outpatient setting for those patients who would otherwise require lifelong hospitalization to obtain a continuous dose of therapeutic protein [49]. Additionally, pegnology allows us to administer biological proteins that normally have a very short half-life and would otherwise require continuous infusions. or multiple daily injections. Therefore by conjugating PEG with proteins we can delay their degradation and enhance their half-life as well as decrease the immunogenicity of the protein. This process facilitates the administration of the bovine form of the enzyme adenosine dearninase to treatment severe combined immunodeficiency disease associated with a deficiency of the enzyme. Table 7.4 lists some of the biopharmaceuticals that were in clinical testing in 1991. The number of these biopharmaceuticals has increased by up to 63%. It is the responsibility of future pharmacists dealing with such products to become knowledgeable in the techniques used to produce these agents, as well as to curate information regarding their uses and adverse effects. Currently, various drug delivery technologies are employed to deliver biopharmaceuticals in the form of dendrimers, liposomes, ceramic nanoparticles, nanocapsules, micelles, hybrid nanodevices and nanospheres.

7.6.4 Biopharmaceutical distribution and preparation

Conventional pharmacists were only responsible for drug preparation and drug distribution. Following the emergence of biopharmaceuticals, there is an urgent need for a professional who has profound knowledge and can look after the all the aspects related to biopharmaceuticals, and counsel patients accordingly [50–52]. Since all biotechnological products are made up of proteins, current biotechnology has taken a new role in providing drug information and pharmaceutical care to patients [50–52]. There should be established guidelines or standard protocols under which the pharmacist's responsibilities in handling biopharmaceuticals and

Table 7.4. Pharmaceutical companies researching drug delivery technologies [88].

Technology	Manufacturer	Status
ViaDerm, a disposable microelectrode array and a patch containing a drug	TransPharma Medical (Yehud, Israel)	Marketed
STEALTH liposomal technology	Alza (Mountain View, CA)	Marketed in USA
Spheramine, dopamine secreting retinal pigmented epithelial cells on gelatin microcarriers	Titan Pharmaceuticals (San Francisco, CA)	Phase IIb
PEGylation technology: PEGASYS (peginterferon α-2a)	Nektar Therapeutics (San Carlos, CA)	Approved by the Food and Drug Administration (FDA), marketed in Europe
PEG-introne for the treatment of hepatitis C	Enzon (Bridgewater, NJ)	Approved by the FDA in August 2001
Microfabricated chips	Micro CHIPS (Bedford, MA)	Preclinical
Low molecular weight polymers are attached to insulin to create drug–polymer conjugates	Nobex (Research Triangle Park, NC)	Phase II
Intraject technology: disposable needle-free delivery system, interferon-a	Aradigm Corporation (Hayward, CA)	Phase I
In vivo gene transfer to skin: DNA vaccines, hormones, proteins	Gentronics Biomedical (San Diego, CA)	Preclinical data
Implantable devices containing cells that secrete therapeutic products	Neurotech (Paris, France)	Phase I

monitoring patients under biopharmaceutical treatment are described. All of these roles are very important as more and more biopharmaceuticals are becoming available for use. Such responsibilities will allow the profession to expand its role in clinical pharmacy services as well as drug information. However, to achieve this, the pharmacist or relevant professional should have the skills and profound knowledge to understand the latest innovations in the emerging area of biopharmaceuticals. This will certainly require upgrading and expansion of contemporary pharmacy services to fulfill the demands imposed by these unique pharmaceuticals.

7.6.5 Distribution

The distribution of drugs or therapeutic proteins is again an important factor, since in most cases the tissue distribution of therapeutic proteins is limited. This is because of the size of these molecules, which is in contrast to smaller molecule drugs that tend to have higher tissue penetration [50–52]. Moreover, factors such as the production process (which may affect post-translational modifications, such as glycosylation), the route of administration (e.g. intravenous versus subcutaneous formulation), physical and chemical properties (e.g. shape and charge) and binding properties (e.g. receptor-mediated uptake) also affect the tissue distribution of a therapeutic protein. Manipulation of these factors through rational design can allow better penetration properties for a biotherapeutic molecule. For example, rational design analysis of factors such as binding affinity and molecular size on tumor targeting was conducted to direct the design of new therapeutic protein drugs [50–54]. However, in contrast to the general administration of therapeutic proteins in tissues, the target-specific delivery of therapeutics is still a demanding, yet very attractive, area for pharmaceutical research. According to many reports, the volume of distribution (after intravenous administration) of large therapeutic proteins, such as mAbs, is close to the plasma volume, suggesting limited distribution into tissues. Findings related to the tissue distribution of radio labeled mAbs indicated that various tissues are exposed to mAbs, but at lower concentrations than typically seen in systemic circulation [50–54]. Despite various, limitations such as limited tissue penetration, large biotherapeutics (such as mAbs) often do have potential even in cases when the site of action is believed to be the tissue. This may indicate the possibility of designing a therapeutic regimen in which tissue exposure is adequate to modulate the target at the site of action. Once these therapeutic proteins reach the blood circulation across capillary endothelial cells and then into tissues, the uptake mechanism commences. The uptake of these therapeutic proteins into cells is carried out by via receptor-mediated transporters (e.g. Fc receptors, often leading to recycling of the molecule) or other internalization processes, such as endocytosis or pinocytosis (often leading to degradation of the molecule). In blood circulation, each biopharmaceutical/therapeutic protein has its own retention time and acquires its own concentration, e.g. high drug concentrations in the kidney and liver have been reported for peptides, low molecular proteins and oligonucleotides. Such a concentration of therapeutic proteins can be achieved in targeted tissue, e.g. a target-mediated tissue distribution of mAbs has been reported [50–54]. After its uptake in tissues, the therapeutic protein undergoes metabolism/catabolism before the remnants of the molecules are excreted from the body. However, in the case of smaller peptides and amino acid degradants, they are recycled for synthesis into other proteins in the body. A list of subcutaneously administered biopharmaceuticals is provided in table 7.5.

7.6.6 Excretion

The pharmacist should be aware of the whole excretion profile of the patient. This will be helpful in determining the renal excretion capacity of the individual under examination. Renal excretion plays an important role in the elimination of protein degradation products and low molecular weight biopharmaceuticals. The whole process of renal filtration, transport and metabolism of low molecular weight proteins has been described well in the literature [50–54]. It has been reported that

Table 7.5. Biopharmaceuticals administered subcutaneously [50–56].

Description (trade name(s))	Mol. Wt. (kDa)	Percentage bioavailability
Anti-p40 mAb (Ustekinumab, Stelara)	150	24%–95% (human), 97% (monkey)
GH (Somatropin, Nutropin)	22	81% (human)
IL-1 inhibitor, fusion protein (Rilonacept, Arcalyst)	251	43% (human), 70% (monkey), 60% (rat), 78% (mouse)
Amylin analog (Pramlintide acetate, Symlin)	3.95	30%–40% (human)
PEG-GH (Pegvisomant, Somavert)	42, 47 and 52[3]	49%–65% (human), 70%–81% (monkey), 45%–73% (mouse)
Cytokine variant (PEG-IFNα-2b, PEG-Intron)	31	57%–89% (monkey), 43%–51% (rat)
Cytokine variant(PEG-IFNα-2a, Pegasys)	60	61%–80% (human)
PEG-G-CSF (Pegfilgrastim, Neulasta)	39	49%–68% (monkey), < 10% to 30% (rat)
Anti-IgE mAb (Omalizumab, Xolair)	149	53%–71% (human), 64%–104% (monkey), 90% (mouse)
IGF-1 (Mecasermin, Increlex)	7.65	100% (human), 47% (rabbit), 38%–57% (rat)
Insulin analog (Insulin lispro, Humalog)	5.81	55%–77% (human)
Insulin analog (Insulin glulisine, Apidra)	5.82	70% (human), 42% (dog), 96% (rat)
Insulin analog (Insulin glargine, Lantus)	6.06	Precipitates in skin-slow uptake in human, dog and rat
IFNβ-1b (Cytokine, Betaseron)	18.5	50% (human), 31%–44% (monkey)
IFNβ-1a (Cytokine, Rebif)	22.5	6%–62% (human), 12%–38% (monkey), 16% (rat)
Anti-TNF mAb (Golimumab, Simponi)	150	53% (human), 77% (monkey)
TNF receptor-Fc-IgG1 fusion protein (Etanercept, Enbrel)	150	76% (human), 73% (monkey), 58% (mouse)
PEG-anti-TNFα Fab fragment (Certolizumab pegol, Cimzia)	91	76%–88% (human), 24%–34% (rat)
Anti-IL-1β mAb (Canakinumab, Ilaris)	145	63%–67% (human), 60% (monkey)
LH-releasing hormone analog (Buserelin acetate, Suprefact)	1.30	70% (human)
Anti-TNF mAbA (Dalimumab, Humira)	148	64% (human), 96% (monkey)

LH: luteinizing hormone; GH: growth hormone; MW: molecular weight; Tox: toxicology (including safety pharmacology); Pharm: pharmacology; SC: subcutaneous; PEG: polyethylene glycol; IFN: interferon; PK: pharmacokinetics; INN: international nonproprietary name; TNF: tumor necrosis factor; IL: interleukin; BAN: British approved name.

proteins are hindered at the glomerular filter in proportion to their molecular size, structure and net charge. However, the mechanisms of reabsorption of certain biopharmaceuticals, such as peptides and proteins in the kidney, need further research. Reports also confirmed that excretion of some very small proteins, e.g. when radio labeled mAbs or Fc fusion proteins were administered in animals, a majority of the radioactive dose was recovered in the urine [50–54], whereas gel electrophoresis (SDS-PAGE) analysis indicated that the radioactive materials in urine were associated with low molecular fragments, suggesting that the excretion of the intact parent drug was negligible. Similarly biliary excretion of therapeutic proteins, such as insulin and epidermal growth factor, has been reported and it was found that proteins were subjected to degradation in the liver, and the degradants were subsequently excreted into bile. Moreover, it has also been suggested that plasma protein binding plays an important role in the tissue distribution of several of biological therapeutics, resulting in altered excretion profiles. Oligomeric lipophilicity backbone alteration is used to prolong the *in vivo* half-life by increasing plasma protein binding in order to reduce the renal excretion [50–54].

7.6.7 Metabolism/catabolism

During metabolism/catabolism, therapeutic proteins are removed from circulation or interstitial fluid via several pathways: target-mediated clearance, fusion protein Fcγ receptor-mediated clearance, nonspecific endocytosis, formation of immune-complexes followed by complement- or Fc receptor-mediated clearance mechanisms, and degradation by proteolysis. In addition to these biological processes proteolysis occurs widely in the body. However, the kinetics and mechanistic details of proteolysis are poorly understood, in particular for large therapeutic proteins such as mAbs. Several *in vitro* and *in vivo* assays can be performed to determine the selection of proteins in discovery research, however, their correlations still remain to be established. For peptides, *in vitro* incubations with plasma, liver and kidney homogenates have been used to facilitate the selection of proteins in discovery research. After the uptake of biotherapeutics into cells, they can be metabolized to peptides or amino acids. This step usually takes place in circulation, carried out by circulating phagocytic cells or by their target antigen-containing cells, or may occur in tissues by various cells. For a therapeutic molecule which is present with an Fc (including therapeutic mAbs, endogenous Abs and fusion proteins), conjugation of the Fc domain to Fc gamma-receptors can result in internalization and subsequent degradation by lysosomes in the reticuloendothelial system (RES) [50–54].

7.6.8 Anti-drug antibodies

Almost all biopharmaceuticals are immunogenic and may induce anti-drug antibodies (ADAs). They may affect patient safety and treatment efficacy. ADAs may induce unwanted side effects, in particular for biotechnology derived pharmaceuticals such as therapeutic antibodies and growth factors. It has been observed that the nonclinical and clinical programming of ADA results in significant changes in toxicology, pharmacokinetics and efficacy. These effects result from the generation of drug induced (neutralizing) auto-antibodies against, e.g. Erythropoietin (EPO) and insulin, and can be responsible for allergic reactions or even anaphylactic shock. Determination of ADAs can be achieved by screening assays (ELISA, bridging ELISA, cytokine profiling), confirmation assays (determination of specificity) and characterization assays (class/isotype of antibodies). If the pharmacokinetics of the active ingredients are available, then it possible to correlate it with the effects induced by ADAs. When biotherapeutics are studied under nonclinical and clinical programs, it has been observed that most therapeutic proteins may suffer from immunogenicity, specifically formation of ADAs. There are various intrinsic and extrinsic factors that contribute to the ability of a therapeutic protein to elicit ADA production. Intrinsic factors affecting immunogenicity are post-translational modification (glycosylation, oxidation), tertiary structure (including aggregation propensity) and protein sequence (including similarity to endogenous proteins and the presence of T and B cell epitopes). Extrinsic factors include the type of formulation (which may affect aggregation), the production process (which may affect both aggregation and post-translational modifications), subject characteristics (disease population, inflammation status, concomitant medications), route, dose, impurities and drug pharmacology (specifically related to immunosuppression) [50–54]. All these factors are considerable in contributing to the changes in ADA response which are observed across the species, subjects and biological modalities.

7.6.9 Glycosylation

According to reports, endogenous and therapeutic proteins can experience from glycosylation, most frequently at asparagine residues ('N-linked') and at serine or threonine residues ('O-linked'). This is the most common, complex and heterogeneric post-translational modification. It has been suggested that inter- and intra-product heterogeneity in glycosylation profile can arise from the variability in glycan type. Moreover, structural features, such as the degree of branching, the site of attachment, the degree of occupancy, and the production system and conditions (such cell type, cell culture medium and purification process) also affect inter- and intra-product heterogeneity. It has been discovered that glycosylation of proteins is important from the absorption, distribution, metabolism and excretion (ADME) and efficacy standpoints, because improperly glycosylated proteins, whether endogenous or exogenously produced biotherapeutics, may be rapidly cleared from circulation by specific receptor based mechanisms, such as high mannose receptors or asialoglyco protein receptors, and because glycosylation may directly affect the biological activity of a biotherapeutic. Therefore, it is very clear that for various approved protein drugs, clinical efficacy depends on proper glycosylation [50–54, 69].

7.6.10 Toxicology

Toxicity evaluation of biotherapeutics and the methods to measure toxicity have been refined. In some cases no toxic effects may be seen, whereas in some cases the toxicity seen can be linked to target-mediated effects. Target-mediated effects

Table 7.6. Examples of on-target pharmacological effects that can be undesirable [50–56].

Examples of exaggerated pharmacology	Action	Undesirable affects
Tumor necrosis factor-α (TNF) inhibitors	Used to treat inflammatory and autoimmune diseases, including rheumatoid arthritis, psoriasis, inflammatory bowel disease and multiple sclerosis.	Infections related to immunosuppression reported in some patients, associated with latent viral or bacterial infections, including *Mycobactrium tuberculosis*, atypical mycobacterial infections, hepatitis B and John Cunningham virus.
B cell depletion therapies	Used for the treatment of B cell tumors, and for inflammatory and autoimmune diseases.	Hypogammaglobulinemia and impaired B-cell reconstitution.
Erythropoiesis	Stimulating agents such as erythropoietins.	At higher doses can cause polycythemia, chronic blood hyperviscosity, vascular stasis, thromboses, increased peripheral resistance and hypertension, which can be fatal.

sometimes produce undesirable effects that are considered to be the result of exaggerated pharmacology, however, they may not be considered to represent primary toxicity [50–56]. Examples of biopharmaceuticals with undesirable effects are mentioned in table 7.6.

7.7 Protein crystals for the delivery of biopharmaceuticals

Recombinant insulin for biopharmaceutical applications is known to be the most successful product in modern biotechnology. This product was approved in 2002 and has become the first crystalline protein to be approved for therapeutic use. Over the last few years, approximately 150 biopharmaceuticals have gained marketing consent. However, only a few successful products such as crystalline insulin protein are still achieving high success rates and have become established products on the pharmaceutical market [57]. With such promising examples, several researchers realized and started working on the efficacy testing and engineering of protein molecules, and crystalline protein production and characterization. The advancements achieved by these researchers triggered the development of crystalline proteins for the treatment of various acute conditions, such as cancer, viral disease and cardiovascular disease [57]. Additionally, such crystalline proteins can also be employed to treat several chronic conditions, such as hemophilia, arthritis, diabetes, psoriasis, Crohn's disease and growth hormone deficiency. Poor stability and limited delivery options are the major challenges for the development of all biopharmaceuticals. Recent findings explored the utilization of protein crystals and their significant advantages in the delivery of biopharmaceuticals for controlled release, low viscosity formulations and achieving high concentration. Such crystalline proteins can be utilized for the delivery of biopharmaceuticals and may help in developing more advanced protein crystallization based technologies [57].

7.8 Patient care, education, training and follow-up

Since all biopharmaceuticals are made up of proteins and may cause serious side effects, patient education will be a very important part of dispensing biopharmaceuticals. These biologically active proteins cannot be delivered orally since they would rapidly be broken down/cleaved by the acidic and enzymatic environment of the gastrointestinal tract. Almost all biopharmaceuticals currently approved in the United States are administered by injection, therefore special training programs, information sources, and actual demonstrations of the proper technique for professionals and patients dealing with biopharmaceuticals are required. Such information can be circulated by pharmaceutical companies or any manufacturer in the form of videos to help instruct the patient on how to self-administer these agents. For the administration of biopharmaceuticals when the patient is not able to self-administer it is also beneficial to educate other members of the patient's family. For example, a recent report established the standard operating procedure for low molecular weight heparin therapy, which suggests the pharmacist's involvement in managing subcutaneous/intramuscular injection therapies [58]. Such education or training programs help to provide comfort to the patient when he or she is administering the drug at home. Special authorities, teams or departments including trained professionals can be organized by government officials for patient counseling, which will further require pharmacies to dedicate an area designed specifically for patient education. In addition to these active programs, written instructions will also help to strengthen the learning from verbal instructions and demonstrations. Follow-up can also be performed for outpatients regarding their history, complications and any specific contraindications related to the biopharmaceuticals.

7.9 Physical factor assessment (related to the serious medical history of the patient)

It is essential to evaluate the physical factors related to the serious medical history of the patient, e.g. cancer and certain neurological disorders. In such a case, pharmacists must be trained to evaluate a patient's aptitude to comprehend the material presented (administering the biopharmaceutical) [59]. Certain physical factors such as poor eyesight or co-morbid states (Parkinson's disease) are also considered, because patients suffering from these conditions would be incapable of self-administration and preparation of an injection on their own. For these patients, the drug should be supplied in prefilled syringes. This would make the procedure more comfortable for the patient and reduces the scope of dosing errors [59].

7.10 Emotional and environmental factor assessment

Various other factors such as patient anxiety level must be evaluated before designing the dosage regimen for a patient undergoing biopharmaceutical therapy [60]. Therefore, emotional and environmental factors must be given equal weight as, e.g. age, and evaluated accordingly [60]. To evaluate these factors some important questions should be assessed by the pharmacist or other trained professional, for example:

- Does the patient have a place to properly store the drug once it is taken home?
- Does the patient have the motivation and mental capability to be compliant with therapy?

All these aspects must be assessed to evaluate whether a patient is a good candidate for home therapy or not.

7.11 Knowledge of products and suitable delivery systems

Developments in biotechnology have led to the production of more potent products and their respective delivery systems. The emerging field of biopharmaceuticals and its various developmental achievements should be carefully monitored by pharmacists to upgrade their knowledge, maintain up-to-date guidelines and implement the same during therapy. Since new biopharmaceuticals available on the market have very little information regarding their pharmacokinetics, adverse effects and action at the molecular level, these new agents may require a level of care never before seen in ambulatory care pharmacies [60–63]. The pharmacist should also be familiar with the safety, storage and stability of the products. Recently, chitosan based systems for biopharmaceuticals and delivery procedure targeting strategies have been the focus of attention. Information related to the physicochemical, pharmacokinetic, pharmacodynamic, pharmacological, toxicological, molecular targeting, biocompatibility and biodegradation properties of such systems should be updated in such a way that the safe level of chitosan should be determined before administering treatment to shellfish allergic patients [60–63].

7.12 Preservatives and biopharmaceuticals

Although several preservatives are approved for parenteral use, such as m-cresol, chlorocresol, phenol, benzyl alcohol, methyl paraben, propyl paraben, benzethonium chloride, benzyl alkynium chloride, thimerosol and chlorobutanol, the chemical and physical properties of these preservatives make the formulation of proteins a challenging task. Potential interaction of preservatives with the protein can also diminish the effectiveness of the preservative. Degradation of preservatives can also produce compounds that interact with proteins. For example, the commonly used preservative benzyl alcohol undergoes oxidation to benzaldehyde which can react with the primary amines of proteins. Long-term instability problems can be addressed by a lyophilized formulation that is combined with a preservative containing diluents, hence shortening the time of exposure of the protein to the preservatives. Most currently approved therapeutic proteins are administered by injection. For intravenous administration the formulation should be clear, free of particles and an iso-osmotic solution with a pH close to 7.4. Suspensions are allowed for subcutaneous or intramuscular tissue and to avoid extremes of pH. Excipients are chosen with care, since most excipients, such as citrate, may also cause pain or stinging upon subcutaneous injection. The compatibility of the protein with the material and solvent system needs to be assessed. The protein human relaxin was found to be degraded when administered through topical methyl cellulose gel. Potential degradation caused by stresses incurred during drug delivery also need to be determined. Delivery routes such as pulmonary and transdermal require a delivery device. Thus the protein's compatibility with the material used in the device as well as the stresses that the protein is subjected to during delivery need to be evaluated, e.g. rhDNase (Pulmozyme®) for treatment of cystic fibrosis is administered using a disposable plastic nebulizer. Although the Pulmozyme® formulation is stable at 2–8 °C for 2 years, these devices subject the protein to an air–water interface over the entire 20 min delivery period [60–63]. Given the susceptibility of proteins to denature at air–water interfaces and to absorb onto plastic surfaces, the integrity of Pulmozyme® after nebulization had to be assessed. Most of these products are manufactured without preservatives. This presents problems for the patient as well as the pharmacist. Laminar flow hoods will become a common fixture in any pharmacy that becomes involved with biotechnology products [60, 61].

7.13 Dosing schedules

The dosing schedules of some of these agents may be puzzling to some patients, since the administration schedule is dependent on indications of the patient, e.g. interferon may be administered daily or intermittently during the week, depending on the patient indication. Similarly, colony-stimulating factors may be administered for seven to ten days following a course of chemotherapy [42, 60, 64]. The best procedure to ensure accurate dosing and patient compliance is to supply a calendar that indicates the dates on which to administer medication. This can ensure patient compliance as well as patient understanding, but there also needs to a commitment to have a pharmacist available and willing to assist patients whenever questions or problems arise. Most biotechnology based drugs must be formulated in dosage forms suitable for human administration in a safe and effective manner. Specific dosage forms with different concentrations and excipients are formulated and designed with special consideration of the route, amount of defined indications and treatment conditions [42, 60, 64]. Reports have revealed that no protein drug to date has been formulated for oral administration and most protein drugs are administered by injection. Parenteral administration encompasses intracardiac, intraspinal, intrathecal, intramuscular, intravenous, intra-arterial, intradermal subcutaneous, intrasynovial and intracutaneous injection, and injection directly into a dermal region. Unlike the other formulation route, the parenteral route of administration requires a much higher standard of purity and sterility than oral administration [42, 60, 64]. It may also require trained personnel to ensure proper administration, which adds to the cost of the treatment. Products intended for subcutaneous injection, however, can be self-administered or administered by a home care provider. As mentioned above, recombinant DNase is given by inhalation aerosol to reduce the viscosity of mucus in the lungs of patients with cystic fibrosis. A platelet derived growth factor in the form of a gel is administered topically for wound healing. Several hormones and peptides are administered through the intranasal route [42, 60, 64]. A solution of the antisense drug used to treat cytomegalovirus retinitis is injected directly into the eye. Dosing schedules are designed on the basis of the patient condition (physical and mental) and route of administration.

7.14 Education regarding adverse effects and prevention strategies for most biopharmaceuticals

Current biopharmaceuticals have important potential for treating a variety of chronic and sometimes life-threatening medical problems. Conventional drugs do not have specified actions, but biopharmaceuticals act specifically, which might also influence their safety profile. Biopharmaceuticals currently suffer from various problems, such as a high potential for immunogenicity, complex production processes, limited predictability of preclinical to clinical data, and adverse events, which are often be related to exaggerated pharmacology. Limitations such as limited predictability of preclinical to clinical data and the known limitations of randomized controlled trials results in limited knowledge of the safety profile. These factors influence their approval at the time of clinical trials which underlines the need for pharmacovigilance [60, 65]. Owing to the different properties of biopharmaceuticals and their special characteristics, pharmacovigilance surveys are essential for biopharmaceuticals, and might differ from those required for small molecules. For pharmacovigilance evaluation, the spontaneous reporting of adverse drug reactions is required for the detection of new, rare and/or serious adverse drug reactions [60, 65].

The importance of educating patients regarding adverse effects is still debated, since there are both advantages and disadvantages to such a training program. Therefore, the deployment of trained professionals should be strongly promoted for this purpose [60, 65]. Most biopharmaceuticals can induce allergic reactions, fatigue, arthralgias, myalgias, fever and headache, which usually pass with time. Special monitoring should be focused on serious allergic reactions, in particular in order to forewarn patients. One of the important roles of the pharmacist will be to educate or train patients on how to minimize these adverse effects [60, 65]. Previous reports have shown that the combination of some drugs effectively reduces the chances of side effects, e.g. administration of 650 mg acetaminophen at bed-time about 30 min prior to a dose of interferon will help to minimize these problems [60, 65]. Instructing patients regarding the development of tolerance against most adverse effects after continued use of the medication is also important to ensure patient compliance. It is very important to make the patient aware of any consequences of discontinuation of the medicine. The patient should also be advised of the symptoms that may recur if dosages are increased or if the drug is discontinued and later restarted. For approved biopharmaceuticals the development of severe adverse effects is rare, however, these therapeutic agents are biological proteins, so allergic reactions, in particular associated with anaphylaxis, are always a concern. Therefore, patients undergoing therapy should be clearly informed about the signs and symptoms of an allergic reaction and also trained in what to do in the case of such a medical condition [60, 65].

7.15 Treatment with adjuvant

Formulation science is a difficult and often overlooked aspect of the field of biopharmaceuticals. Understanding the relationship between the adjuvant and the antigen helps in determining unexpected immunological responses generated by these complex biopharmaceutical formulations. It is clear that some biopharmaceuticals (e.g. vaccines) will require adjuvant treatment with other pharmaceuticals, e.g. the important role of iron supplementation during the course of epoetin-α therapy [66]. Such supplementation is very important for those medical conditions where the patient is iron deficient, as the epoetin-α administration will otherwise produce less than the desired effect, since iron supplementation is essential for red blood cell formation [66]. Patient refusal of iron supplements is common, but with the assistance of a trained pharmacist the patient can be encouraged to comply so that they receive the maximum benefit from the therapy [66].

7.16 Dosage errors

Biopharmaceutical administration errors are potentially dangerous occurrences that many immunization providers miss. Errors in biopharmaceutical administration are common, but relatively few data are available on why they happen. Dosage errors can happen very easily, e.g. diphtheria, tetanus and pertussis toxoids Tdap (for adolescent and adult use) and DTap (for pediatric use) have similar nonproprietary names and vaccine abbreviations (DTaP and Tdap) which is believed to have contributed to dosage confusion [67]. To prevent errors, separate stock of the pediatric and adult formulations should be kept, shelf-talkers can be used to direct staff to the location of each formulation, and alerts can be placed on the products (e.g. 'Adult' or 'Pediatric') and on automated dispensing cabinet screens, if applicable [67]. Moreover, correct dosage calculations, although a concern when dispensing any pharmaceutical, are of particular concern for biopharmaceuticals. Various biotherapeutics are dosed using different units and may be based on body weight or body surface area. Careful calculations should be performed and cross-checked, preferably by another pharmacist, prior to dispensing the product to the patient. Such procedures will ensure the delivery of the correct dose and minimize the chances of treatment failure or adverse effects due to under- or over-dosing.

7.17 Proper supply

The proper supply factor indicates the convenience of the medical device and mode of administration selected during therapy and the measurement/calculation/maintenance of the exact dose regimen for the patient [68]. It is also essential to learn about the importance of a proper supply of the dose considering the convenience for the patient in the self-administration of drugs. This can be achieved by using a suitable medical device, such as tuberculin syringes or, for doses less than 0.3 ml, the use of a low-dose insulin syringe may be required [68].

7.18 Familiarity with terminology and the biotechnological literature

As mentioned above, there are some similarities between the names and formulations of some biopharmaceuticals, therefore there are some obligations when dealing with these products. The first and most common obligation of a pharmacist is to differentiate between products with similar names. For this, the pharmacist should be familiar with the standard terminology and the biotechnological literature. The pharmacist should be regularly trained to maintain an understanding of the agents, their uses, their adverse effects and their mechanisms of action. Moreover, it is the responsibility of the pharmacist who is dealing with biopharmaceuticals to frequently monitor information sources to keep abreast of continuing developments in the field.

7.19 Dealing with expensive biopharmaceuticals

One of the major obstacles for patients experiencing biopharmaceutical therapy is the cost, since most biopharmaceuticals are very expensive. Even to keep such products at pharmaceutical retail stores would require a very large initial investment, since most biopharmaceuticals are protein based, therefore in addition to the acquisition cost of these agents, procurement facilities are required which would certainly increase the expense on the part of the pharmacy. As a recent example, the acquisition cost of epoetin-α (3000 units three times weekly) for a month's supply is more than 400 US dollars. Similarly, a course for patients undergoing colony-stimulating factor therapy may require an expenditure greater than $2000. Such expensive therapies will prevent some pharmacies from procuring biopharmaceuticals as regular stock items. This will lead to the development of some pharmacies which are efficient in delivering expensive biopharmaceuticals continuously on demand or quickly on an as-needed basis. Currently, many wholesalers and pharmaceutical companies offer such services for biopharmaceuticals that are approved for usage. It is also essential to inform physicians and patients to notify the pharmacy in advance of an anticipated need for these products. Close association and understanding between the pharmacist and physician is required to prescribe biopharmaceuticals and to help in inventory control. In addition, continuous follow-up with patients can alert pharmacies to expected requirements to order more of a particular agent. Giving consideration to these factors will facilitate the pharmacist to keep stock low and thus not tie up funds in the acquisition of these costly agents.

Owing to the increasing demand for and cost of biotechnology based products, legal authorities are now trying to establish certain standards for suppliers and monitor their role in health interventions [70–72]. To supplement decision making, some legal advisers are still relying on cost–utility analyses (CUAs). Over 3300 CUAs had been carried out up to 2012, and these CUAs estimate the resources used (costs) and the health benefits achieved (effects) for an intervention compared to an alternative treatment strategy. The cost is a major issue that impedes the growth of biopharmaceuticals. In 2010, biopharmaceutical spending increased by 6.6%. This value is nearly three times the percentage for other pharmaceuticals. Biopharmaceuticals are the sixth major class of interventions in the Registry—258 (11%) of the 2383 cost–effectiveness analyses (CEA) (reported from 2009 data) [70–72]. Analyzing of Cost-Effectiveness Analysis Registry revealed the cost–utility literature for biopharmaceuticals and compared their value to other health interventions. Of 2,383 studies in the registry through 2009, biopharmaceutical CUAs comprised the sixth largest category of interventions at 11%. According to previous findings, the median cost-effectiveness ratio of biopharmaceuticals was not as good as that of other interventions. Wilson and Newman (2012) explored logistic regression among biopharmaceuticals. This regression suggested the cost-effectiveness of industry-sponsored studies, which clearly showed that products used in the treatment of infectious ailments were considerably more likely to have a favorable CUA. In contrast, the results for cancer and neurological treatments were significantly less favorable [70–72]. The costs involved in the treatment of several diseases using biopharmaceuticals are listed in table 7.7.

7.20 Generic biopharmaceuticals

Cost regulation in the healthcare industry is a major problem. The introduction of generic drugs on the market can cause a decline in the price of prescription drugs. So far, there is no government policy that can regulate generic/brand competition in the biopharmaceutical market [73, 74]. Current research suggests that biopharmaceuticals make up 12% of the prescription drug market. This 12% share is expected to increase significantly due to the developments occurring in this field. The subject of generic biopharmaceutical drugs has become controversial since the patent protection for various biological drugs (filgrastim, Neupogen) with substantial sales expired in 2006 [73, 74]. According to a recent survey, the cost of biopharmaceuticals is significantly higher than that of most chemical drugs. The high cost creates concern for those who pay for the drugs. Biopharmaceutical therapy is known to be one of the most expensive, since the approximate cost of the top three selling biopharmaceuticals epoetin-α (Epogen, $10 000), filgrastim (Neupogen, $15 000) and interferon α-2b (Intron A, $22 000) per patient per year is very high [73, 74]. Other expensive biopharmaceutical therapies, such as imiglucerase (Cerezyme), can cost nearly $200 000 per patient per year [73, 74]. This high cost promotes competition in the market, and improvements in production technology which

Table 7.7. CUAs for certain treatments [70–72].

Intervention	Patient	Cost (US dollars)
Abatacept plus oral disease modifying antirheumatic drugs in patients with moderate to severely active rheumatoid arthritis.	Elderly male/female	\$46 000–\$51 000
Annual HIV screening.	For patients with low to moderate risk	Enhances expenditure and deteriorates health
Daily dialysis of a patient suffering from kidney failure.	60-year-old men	\$6000
Diabetes education and self-management of patients with type 2 diabetes.	Diabetic patients	\$4000
First line bevacizumab in combination with irrotecan and 5-FU/LV in treating metastatic colorectal cancer.	Elderly men	\$130 000–\$180 000
HIV counseling, testing and referral.	For a high risk population	\$44 000
Methotrexate with infliximab in treating rheumatoid arthritis.	Elderly men	\$12 000
Omalizumab in patients with severe persistent allergic asthma.	Male/female adult/ elderly	\$43 000
Preventing esophageal cancer with daily aspirin administration.	55-year-old men	Saves money and recovers health
Ranibizumab for the treatment of subfoveal neovascular macular degeneration.	Elderly men	\$54 000–\$130 000
Screening for osteoporosis.	65-year-old men	\$150 000
Screening heavy smokers with CAT scans.	60-year-old men	\$140 000
Treatment with epoetin-α in women with stage IV breast cancer.	Adult/elderly female	\$17 000
tPA + heparin supplementation to remove blood clots in the lungs.	Patients suffering from right ventricle dysfunction	Enhances expenditure and deteriorates health
Training caregivers for stroke patients.	Patients with a past history of stroke	Saves money and recovers health
Treating rheumatoid arthritis with drugs that slow disease progression.	Elderly male/female	Saves money and recovers health
Treating sciatica with early surgery.	Elderly patients	Saves money and recovers health

Treating spinal stenosis with surgery.	Patients suffering from leg pain and spinal narrowing.	\$90 000
Treatment of multiple sclerosis with interferon β-1b.	Adult/elderly men	\$50 000–\$1 800 000
Using Coumadin (warfarin) for arterial fibrillation.	70-year-old men/women	\$3000
Using an implantable Cardioverter defibrillator to prevent sudden cardiac death.	High risk patients	\$38 000

will decrease the estimated costs. Competition between branded drugs and generic drugs may also help in controlling the costs. Generic biopharmaceuticals are advantageous for patients or consumers who require biotherapeutics for different ailments, such as rheumatoid arthritis, hepatitis C and multiple sclerosis, at an economical price [73, 74]. However, issues such as safety, efficacy, economics, legality and public policy can impede the development of generic biopharmaceuticals. This biogeneric market allows trading of chemical drugs, standard biopharmaceuticals, generic biopharmaceuticals and various alternative therapies, such as some surgical operations which can be carried out to avoid the cost of biopharmaceuticals. The issue of generic biopharmaceuticals was hotly debated after the approval of the generic biopharmaceutical Omnitrope (somatropin with an rDNA origin) for marketing by the US Food and Drug Administration and the European Agency for the Evaluation of Medicinal Products (EMEA) [73, 74]. The proportion of profit gained from follow-on biological drugs may not be as large as that achieved with generic chemical drugs, however, possible investments could be considerable.

7.21 Collaboration between pharmacies, wholesalers and manufacturers

To ensure biopharmaceutical delivery to the patient in a timely manner, it is important to have collaboration between pharmacies, wholesalers and manufacturers. Such collaboration not only allows them to work together, but also relieves the patient from dealing with three parties independently [60, 61]. Such teamwork can be promoted by the development of consignment or banking programs. To achieve such a goal, pharmacies could be permitted to stock biopharmaceuticals (after approval of a procurement facility by the governing body) and they will be paid for them once the products are dispensed [60, 61].

7.22 Reimbursement issues and biopharmaceuticals

Since most biopharmaceuticals are expensive, reimbursement issues will be a matter of concern during the delivery of treatment. For this, the pharmacist should select a proactive approach and develop a strategic plan to deal with this issue. This will be important for any pharmacy that moves into the world of biopharmaceuticals [60, 74–77]. One of the alternative routes is that the supplier can have a huge debt to biopharmaceutical manufacturers for a certain period, until reimbursement from insurance companies or the government is received [60, 74–77]. After the payment of the whole amount, the return often does not meet the cost. It was also observed that under some reimbursement policies, biopharmaceutical administration can be a cost instead of generating a source of revenue. Owing to the high costs and long reimbursement periods some practitioners are now foregoing the costs of vaccines altogether [60, 74–77].

7.23 Third party reimbursement issues with biopharmaceuticals

To deliver biopharmaceuticals on demand, either a large amount of investment can be made to stock these agents or alternatively unique agreements can be generated with wholesalers, manufacturers and agencies accountable for third party reimbursement [78–83]. Most biopharmaceutical industries have set up special programs to help with this issue. Most insurance companies will reject the utilization of biotherapeutics for use outside the labeled instructions [78–83]. Self-administration and utilization of these agents in the home have also been a problem. It has been found that third party programs usually object to and reject reimbursment for epoetin-α, unless it was administered in a hospital or under the supervision of a physician. It is the responsibility of manufacturers, pharmacies and physicians to inform these companies that the self-administration of biopharmaceuticals is a safe and effective means of delivering these medications [78–83]. Biopharmaceutical manufacturers can be asked to provide support for unaffordable biopharmaceutical therapy to those who find it difficult to self-administer and require the assistance of a professional. In such cases, there should be an understanding between the insurer and manufacturer, where the manufacturer provides assurance about the safety of these agents to the insurer, who may then reconsider [78–83].

7.24 Affordable therapy and biopharmaceuticals

It is very important for pharmacists/insurance agents to select the most affordable and effective therapy for patients, in particular when the patient is not in a suitable state to select between different products with different price ranges produced by different manufacturers. For this purpose, some manufacturers have set up protocols to assist patients who are unable to pay for their medication. Here the involvement of pharmacies is required to provide medication to those who are unable to afford the high cost of therapy.

7.25 Information on the right reimbursement policy for biopharmaceuticals

The responsibilities of pharmacists/manufacturers/insurance agents do not end with products only, they are also responsible for providing patients with information regarding the right reimbursement policy [60, 78–83]. There are various reimbursement issues surrounding these products, thus the pharmacist should maintain information regarding the product and its reimbursement issues. Additionally, the pharmacist should counsel patients regarding reimbursement and also work with the patient to determine coverage. Since the medical states of some patients going through such therapies are very critical, assisting them with reimbursement problems can provide an invaluable service [60, 78–83].

7.26 Direct reimbursement of biopharmaceuticals

A primary concern of the pharmacist is to provide the reimbursement as soon as possible to the patient, although this rarely happens and usually it will take time to cover all the official procedures/documentation required during this process [60, 78–83]. This reimbursement process will remain time-consuming unless a quick protocol to speed up the process is developed, or unique agreements with wholesalers and/or pharmaceutical companies are made. Alternatively, there could be a direct reimbursement procedure to the company or wholesaler instead of the pharmacy [60, 78–83]. The role of the pharmacist is very important in establishing a protocol for agreements with either manufacturers, wholesalers or both to facilitate the longer periods before payment. This would give the pharmacy sufficient time to be repaid before compensation is due.

7.27 Dispensing fee reimbursement when dealing with biopharmaceuticals

Some other issues, such as the fees charged during the dispensing of biopharmaceuticals, are also a matter of concern. The involvement of third party payers allows a minimum dispensing fee to be added to the complete purchase cost of the biopharmaceuticals [84]. However, this fee can be restructured according to the time required for patient training and education. By considering the pharmacy's commitment to providing a higher level of pharmaceutical care, third party payers offer different structures of reimbursement [84]. For such purposes there is a requirement for only pharmacists trained by an appropriate body to be eligible for this restructured reimbursement.

7.28 Proper refrigeration of biopharmaceuticals

Most biopharmaceuticals have special instructions for storage that differ from product to product. Owing to the varying degree of instability caused by preservatives, most biopharmaceuticals are manufactured without preservatives. Therefore, for the procurement of these biological proteins optimum refrigeration is required [85]. Freezing facilities or conditions may vary, e.g. some agents require storage in a freezer, whereas others cannot be frozen because the product would be denatured. Basic facilities should not cause any real problems as most pharmacies already have refrigerators [85]. However, temperature monitoring will be more critical with biopharmaceuticals, since some agents are unstable if kept at ambient temperatures for long periods of time, e.g. Sargramostim is useless if kept at room temperature for more than six hours, in contrast, filgrastim remains stable and is safe to use for up to 24 h at room temperature [85, 86]. The stability of filgrastim will only be ensured after inspecting the clarity of the solution and there should be no particulate matter suspended in it [85, 86]. The storage factor is also an issue during the transport of these agents from the pharmacy to the patient's home, in particular in areas where the patient lives many hours travel from from the pharmacy. In such cases, coolers must be provided for patients who have long distances to travel from the pharmacy to their homes [85, 86]. Currently, there are different biotechnological devices accessible for patients with diabetes to carry their insulin with them and still keep the insulin at refrigerator temperatures. These products could be modified for use with other biopharmaceuticals. Frequently used biopharmaceuticals (required in a short time during the course of therapy) could be available to patients to allow them to go on extended trips while continuing therapy [87].

7.29 Stock management of biopharmaceuticals

The maintenance of a stock of these protein based products can also generate other challenges to pharmacists. These protein based medicines have short expiration times in comparison to other medicines that are protected with excipients (which protect the drug better but cause incompatibility with protein based drugs) [60]. Constant monitoring is required by pharmacists to monitor those drugs that have short expiration times. Policies that are introduced by manufacturers need to be updated/modified to allow replacement of expired medication.

7.30 Drug loss due to premature discontinuation of medications

Since biopharmaceutical therapy involves some serious concerns related to expense, intolerable adverse effects or other complications, and therapy is dependent upon the response of the patient, some patients discontinue therapy prematurely [60]. Large doses of medication are required to supply the drug for at least one month, so if the drug is discontinued prematurely by the patient during treatment, a considerable amount of drug could be wasted [60]. Since the agents are very expensive and their availability is limited, it is undesirable to waste such drugs. Therefore pharmacies should coordinate with the patient and supplier to dispense drugs for no longer than two weeks. Pharmacies can develop policies according to the patient's compliance to the initial period of therapy [60]. If the therapy is intolerable (with too many adverse effects) and may require a dosage change or discontinuation of therapy, then pharmacies either follow the developed policy or consult the manufacturer/insurance agent accordingly. The rules and regulations in several countries do not permit the patient to return drugs, thus amendments of such laws could allow unopened vials of these agents to be returned.

7.31 The future pharmacist who can efficiently deal with biopharmaceuticals

With the emergence of biopharmaceuticals there is an urgent need for skilled and proficient professional pharmacists with profound knowledge related to dealing with patients undergoing biopharmaceutical therapy. The future pharmacist can be defined as a pharmacist who can be trained to deal with more advanced products such as biopharmaceuticals, and the time is not too far off when pharmacists could be delivering engineered human genes in ambulatory care settings. With the introduction of more advanced therapies such as gene therapy, the chances of curing various diseases (attributable to alterations, mutations or deletions of a gene) are increased. Suitable gene administration that corrects a defective gene and its associated disorder will provide the ultimate therapy against such ailments. This area of gene therapy requires more advanced and new drug delivery systems for the delivery of the most basic elements of therapy, i.e. proteins, peptides or nucleic acids. Such products require special handling by a skilled professional who is well trained and familiar with all specialized preparation and storage methods as well as sophisticated monitoring systems.

The field of biotechnology has produced an array of products that create several challenges for the pharmacist. Association of the pharmacist with the physician and patient in ambulatory care settings requires the delivery of these products on time. After proper examination, the pharmacist could take a decision to dispense these agents or to refer the patient to another pharmacy. This judgment will have great effects on the entire pharmacy, since this decision is accompanied by other responsibilities, such as patient education, training and follow-up, as well as drug acquisition and reimbursement. The pharmacist will need to keep abreast of current research on the immune system and the biological roles of the proteins being administered.

7.32 Stability issues of biopharmaceuticals

Most biopharmaceuticals are made up of proteins, which can impede their development pathway since there are several factors that affect the structure of proteins. There are various physical and chemical factors that can influence the quality and stability of biopharmaceutical products, mainly after long-term storage, subject to variations in light, temperature, and agitation with shipping and handling. In contrast to general pharmaceuticals, proteins are considerably larger molecular entities with inherent physiochemical complexities, from their primary amino acid sequences through to their higher-order secondary and tertiary structures and, in some cases, quaternary elements such as subunit associations [89]. It has been discovered that certain proteins are glycosylated, and some have other post-translational modifications (e.g. phosphorylation). These processes affect their potential degradation pathways as well as the kinetics of their degradation. Moreover, it has been discovered that certain proteins are characteristically susceptible to minor alterations in solution chemistry, i.e. they remain compositionally and conformationally stable only within a relatively narrow range of pH and osmolarity. In contrast, many require additional supportive formulation components to remain in solution, particularly over time [90]. According to reports, it has been observed that even lyophilized and spray protein products experience degradation [91–93]. Therefore, pharmacists should be aware of all the factors that influence the stability of biopharmaceuticals, so that he/she can instruct/train the patient accordingly. Additionally, the pharmacist should be aware of all the characterization and regulatory procedures for follow-on versions of such biopharmaceutical products (also known as biosimilars) [94–96]. Pharmacists should be aware of the role of analytical technologies in comparing biosimilars to the corresponding reference products, which is attracting substantial interest in establishing the development requirements for biosimilars [94–96]

Table 7.8. Approved biopharmaceuticals for use in the United States.

Generic name	Trade name	Contraindications	Adverse effects
Somatropin	Humatrope	Patients allergic to synthetic growth hormone, respiratory problems, open heart surgery, overweight patients, diabetic retinopathy, sleep apnea, Prader–Willis syndrome.	Headache, joint pain, redness, nausea and vomiting, swelling at the injection site, increased thirst and urination, stiffness, gas, breast swelling.
Somatrem	Protropin	Pregnancy, tumors, allergies.	Fatigue or muscle pain, headache.
Sargramostim	Leukine	Cancer, lung/heart problems, allergies, kidney disease, liver disease, fluid retention, other blood disorders, chemotherapy.	Headache, chills, aching in the bones and muscles (taking a non-aspirin pain reliever such as acetaminophen can help).
Pegademase bovine	Adagen	Thrombocytopenia; should not be used if the condition is severe.	Signs of infection such as sore throat, allergic reaction, fever or congestion, anemia.
Muromonab-CD3	Orthoclone OKT	Arterial hypertension, epilepsy, pregnancy, lactation, allergies to mouse proteins, uncompensated heart failure.	Cytokine syndrome (glucocorticoids, acetaminophen and diphenhydramine can be given).
Interferon γ-lb	Actimmune	Hepatic decompensation, autoimmune hepatitis, hypersensitivity reactions, pregnancy, hemoglobin opathies.	Flu-like symptoms (managed with analgesics and antihistamines), depression and suicide, anorexia.
Interferon α-n3	Alferon N	Hypersensitivity to human proteins or any component of the product, anaphylactic sensitivity to mouse immunoglobulin (IgG), egg protein or neomycin.	COPD, diabetes, allergic reactions, heart disease, congestive heart failure, angina, bone marrow suppression, blood clotting, seizure disorders.
Interferon α-2b	Intron A	Pregnancy, hemoglobin opathies, hypersensitivity, autoimmune hepatitis.	Mild flu-like symptoms, constipation, diarrhea, loss of appetite, dizziness, severe allergic reactions.
Interferon α-2a	Roferon-A	Hepatitis, hypersensitivity, hepatic decompensation, autoimmune conditions.	Bone marrow depression, depressive illness, suicidal behavior, insomnia, anxiety, GI hemorrhage, flu-like symptoms, triglyceride levels.

Human insulin	Humulin	Hypoglycaemia, severe allergy to bovine insulin or porcine insulin.	Hypoglycemia, weight gain, loss of fatty tissues, allergic reactions, upper respiratory tract problems.
Hepatitis B vaccine	Engerix-B, Recombivax HB	Leukemia, hemodialysis, pregnancy, low platelets, HIV, lymphoma, hemophilia.	Fever, headache, loss of appetite, fainting right after injection, nausea pain/soreness/redness/swelling at the injection site, tiredness, sore throat, diarrhea.
Filgrastim	Neupogen	Sickle cell/heart disease, spleen problems, allergic/skin reaction, blood disorders, pregnancy.	Allergic reactions, nosebleeds, aching in the bones and muscles, stomach/abdominal pain.
Epoetin-α	Epogen Procrit	Pediatric patients, uncontrolled hypertension, allergic reactions, neonates, pure red cell aplasia, infants, pregnancy.	Rash, itching/swelling, dizziness, body aches, diarrhea, irritation, trouble breathing, headache, high blood pressure, kidney failure.
Alteplase	Activase	Pregnancy, bleeding disorders, recent injury or surgery, liver disease, high blood pressure, endocarditis, allergies.	Vomiting, nausea, skin rash, dizziness, mild fever, easy bruising or bleeding, rapid or abnormal heartbeat, trouble breathing, chest pain.
Aidesleukin	Proleukin	Myocardial infarction, cardiac tamponade, renal failure, cardiac arrhythmias, toxic psychosis, seizures, ventricular tachycardia, GI bleeding, bowel schemia.	Loss of appetite, behavior and skin changes, anemia, tiredness and weakness, fluid retention.

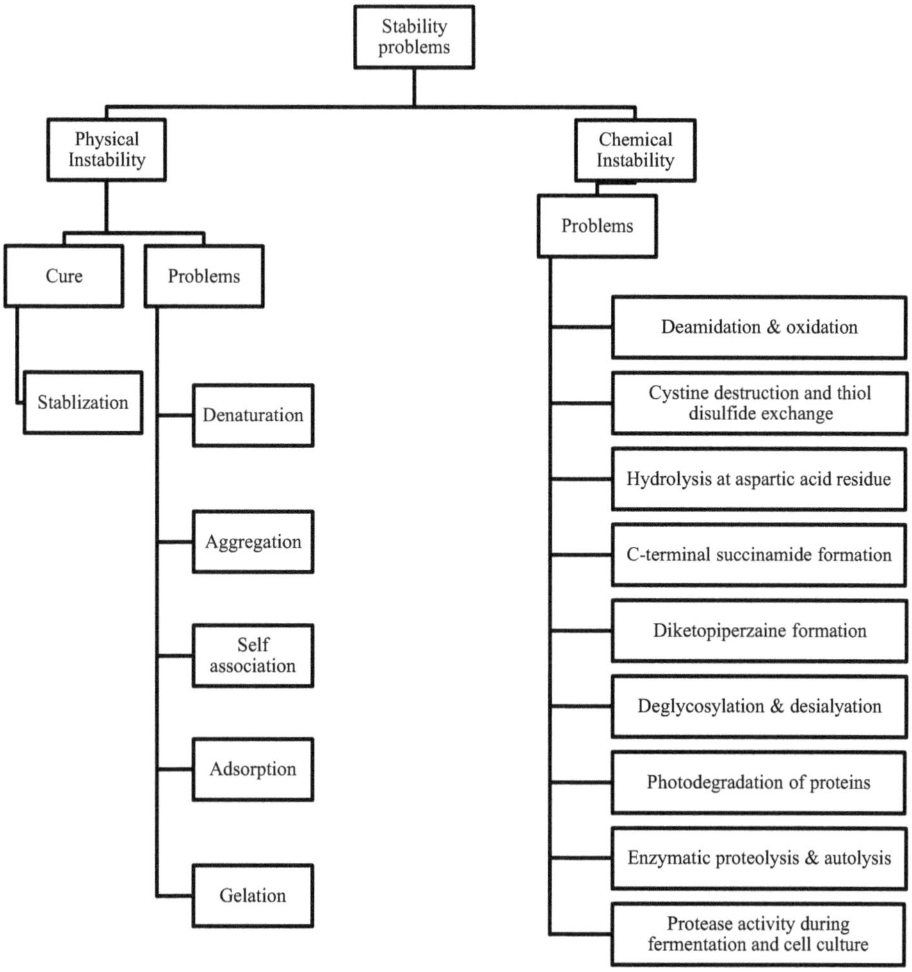

Figure 7.5. Factors that affect the stability of proteins.

A list of the approved biopharmaceuticals for use in the United States is provided in table 7.8. Factors that affect the stability of proteins are listed in figure 7.5.

Various biotechnology based products, their function and their manufacturer are listed in table 7.9. These biopharmaceuticals can be utilized to treat various complications such as:

- Treatment of anemia associated with chronic renal failure.
- Severe combined immunodeficiency disease associated with a deficiency of adenosine deammase.
- Reversal of acute kidney transplant rejection.
- Patients on dialysis and not on dialysis.
- Management of chronic granulomatous disease.
- Human growth hormone deficiency in children.
- Hepatitis B and hepatitis C.

Table 7.9. Various biotechnology based products.

Biopharmaceutical	Product	Function	Marketed product and manufacturer
Clotting factors	Factor VIIa	Causes blood to clot in the coagulation cascade.	Bioclate—rh-factor VIII produced in CHO cells (Centeon, 1993); Kogenate—rh-factor VIII produced in BHK cells (Bayer, 1993), also sold as Helixate by Centeon; Recombinate—rh-factor VIII produced in an animal cell line (Baxter Healthcare/Genetics Institute, 1992).
	Factor IX	Treating hemophilia B, most common type of hemophilia, or Christmas disease, due to deficiency of factor IX.	
	Anti-hemophilic factor (factor VIII)	Blood clotting protein, known as the anti-hemophilic factor, encoded by the F8 gene. Defects in this gene result in hemophilia A.	
Colony-stimulating factors (CSF)	Macrophage CSF	Cells to proliferate and differentiate into a specific kind of blood cell in bone marrow stimulation.	Neupogen—filgrastim, rG-CSF produced in *E. coli* (Amgen, 1991) in chemotherapy-induced neutropenia
Dismutases	Superoxide dismutase	Enzymes that catalyze the dismutation of superoxide into oxygen and hydrogen peroxide. Control antioxidant defense in nearly all cells exposed to oxygen, in particular in respiratory disorders.	Dismutases (Xiamen Runcheng Pharma Group, China)
Erythropoietins	Epoetin-β	Glycoprotein hormone that controls erythropoiesis, or red blood cell production in anemia.	Epogen—rh-EPO produced in a mammalian cell line (Amgen, 1989); Procrit—rh-EPO produced in a mammalian cell line (Ortho Biotech, 1990)

(Continued)

Growth factors	Epidermal growth factor (EGF)	EGF is a growth factor that stimulates cell growth, proliferation and differentiation by binding to its receptor.	Recombinant human epidermal growth factor, rh-EGF, rh-EGF, EGF, recombinant human keratinocyte growth factor 2KGF-2 (Hunan Weih-Lugu Biotech)
	Fibroblast growth factor (FGFs)	FGFs are involved in angiogenesis, wound healing and embryonic development.	
	Insulin-like growth factor (IGFs)	IGFs are proteins with a close sequence similarity to insulin.	
	Platelet derived growth factor (PDGF)	PDGF regulates cell growth and division, in particular in blood vessel formation (angiogenesis) and the growth of blood vessels from already-existing blood vessel tissue (wound healing).	
Growth hormone (GH)	Somatotropin (STH)	GH stimulates growth, cell reproduction and regeneration in humans, clinically used for short stature and multiple sclerosis. STH is produced naturally in animals and it can be produced by rDNA technology.	Protropin—rh-GH, produced in *E. coli* (Genentech, 1985); Humatrope—rh-GH produced in *E. coli* (Eli Lilly, 1987); Nutropin—rh-GH produced in *E. coli* (Genentech, 1994); BioTropin—rh-GH (Biotechnology General, 1995); Genotropin—rh-GH produced in *E. coli* (Pharmacia and Upjohn, 1995); Norditropin—rh-GH (Novo Nordisk, 1995)
Somatostatin—growth hormone-inhibiting hormone (GHIH),	Somnidren GH	Regulates the endocrine system and affects neurotransmission and cell proliferation via interaction with G-protein-coupled somatostatin	Somatostatin (Genetech, 1977), Sandostatin, Sandoz, antidiarrheal-synthetic hormone

somatotropin release-inhibiting factor (SRIF) or somatotropin release-inhibiting hormone (SRIH)		receptors, and plays a major role in glucose regulation.	
Insulin	Humulin, proinsulin, triproamylin	Insulin regulates carbohydrate and fat metabolism by allowing the cells in the liver, muscle and fat tissue to utilize glucose from the blood and further store it as glycogen inside these tissues.	Humulin—rh-insulin produced in *E. coli* (Eli Lilly, 1982); Novolin—rh-insulin (Novo Nordisk, 1991; also Wockhardt, the world's fourth and Asia's largest producer of human insulin).
Interferons (named after their first discovered function in interfering with viral replication)	Type I IFN, type II IFN, and type III IFN	Signaling proteins (cytokines), released by host cells in response to the presence of pathogens. In a typical scenario, a virus-infected cell will release interferons causing nearby cells to heighten their anti-viral defenses.	Intron A—rIFN-a-2b produced in *E. coli* (Hoffman-La-Roche, 1986) for hepatitis; Roferon-A—rh-IFN-a-2a, produced in *E. coli* (Hoffman-La-Roche, 1986) for hairy cell leukemia; Actimmune—rh-IFN-a-1b produced in *E. coli* (1990) for chronic granulomatous disease.
Interleukins (IL), produced by leukocytes and act on leukocytes	PEG IL-2IL-l αIL-l βIL-3IL-4IL-6	IL are cytokines that regulate the immune system and deficiency of IL causes autoimmune diseases or immune deficiency. The majority of interleukins are synthesized by helper CD4 + T lymphocytes, monocytes, macrophages, and endothelial cells.	Recombinant human interleukin-2/4, rHuIL-2 rHuIL-4 (Hunan Weih-Lugu Biotech)

(Continued)

Monoclonal antibodies (MAb or MoAb)	Murine MoAb, chimeric MoAb, humanized MoAb, human MoAb	MoAbs are monospecific antibodies that are the same because they are made by identical immune cells that are all clones of a unique parent cell. Monoclonal antibodies have monovalent affinity, in that they bind to the same epitope. They are used in the treatment of cancer and autoimmune disease	Indimacis 125—Igovomab, murine MAb fragment (CIS Bio, 1996); Tecnemab KI—murine MAb fragment (Sorin, 1996); LeukoScan—Sulesomab, murine MAb fragment (Immunomedics, 1997); Zevalin—Ibritumomab (IDEC pharmaceuticals, 2002); Bexxar—tositumomab (Corixa/GlaxoSmithKline, 2003); Orthoclone OKT3—Muromomab CD3, murine MAb (Ortho Biotech, 1986); OncoScint CR/OV—Satumomab Pendetide, murine MAb (Cytogen, 1992); CEA-scan—Arcitumomab, murine MAb fragment (Immunomedics, 1996)
Recombinant soluble CD4s	CD4-IgGrsCD4	CD4 (cluster of differentiation 4) is a co-receptor glycoprotein found on the surface of immune cells that assists the T cell receptor (TCR) with an antigen-presenting cell.	sCD4 recombinant protein
Tissue plasminogen activators (abbreviated tPA or PLAT)	Tissue plasminogen activator (Fibrinolytic)	tPA is a protein involved in the breakdown of blood clots.	Activase—Alteplase, rh-tPA produced in CHO cells (Genentech, 1987), for acute myocardial infarction; Retavase—tPA analogue (Boehringer-Manheim/Centocor); Ecokinase (Galenus Mannheim); Rapilysin (Boheringer-Manheim); TNKase—tPA analogue (Genentech/Schering-Plough)

Streptokinase (SK)	Acylated plasminogen SK activator complex, recombinant fusion SK, modified SK with reduced immunogenicity, mutant SK with improved stability	SK was the first FDA approved thrombolytic drug; it does not have any proteolytic activity as of itself, instead forms a 1:1 stochiometric complex with a plasminogen or a plasmin molecule.	BBT (Biotech); Heberkinase (Heber Biotec); Streptonase (Kyung Dong Pharm); Streptokinase TTK (BBT Biotech); Eskinase (BBT Biotech); Recombinant streptokinase (Shanghai SIIC SMU Biotech); STPase (Cadila Pharmaceuticals); Kabikinase (Pharmacia Upjohn)
Hirudin	Desirudin, a recombinant form of hirudin (fibrinolytic)	Hirudin is a naturally occurring peptide in that has a blood anticoagulant property.	Recombinant proteins, rHuGM-CSF, hirudin, r-hirudin (Hunan Weih-Lugu Biotech)
Tumor necrosis factors (or the TNF-family)	Tumor necrosis factor and lymphotoxin-α are two major types	TNFs refer to cytokines whose family can cause cell death (apoptosis).	Enbrel®—etanercept (Immunex Corporation)
Vaccines	AIDS vaccine, herpes vaccine, malaria vaccine, melanoma vaccine	A vaccine is a biological preparation that improves immunity to a particular disease.	Hepatitis B surface antigen vaccine—CHO cells, rhHBsAG (Hunan Weih-Lugu Biotech); Shanta Biotech was the first Indian company to get WHO certification for hepatitis B vaccine; Wockhardt had the first WHO-GMP certified hepatitis B vaccine
Follicle-stimulating hormone (rh-FSH)	rh-FSH	Synthesized and secreted by gonadotropes of the anterior pituitary gland and regulates the development, growth, pubertal maturation and reproductive processes of the body.	Gonal F—rh-FSH produced in CHO cells (Serono, 1995) for anovulation and superovulation.

(*Continued*)

α1 antitrypsin (AAT)	Zemaira®	AAT deficiency is a condition in which the body does not make enough of a protein that protects the lungs and liver from damage.	Four augmentation therapies are manufactured by the following companies: Grifols, Baxter Therapeutics, CSL-Behring, Kamada
Atrial natriuretic peptide (ANP)	Atrial natriuretic factor (ANF), atrial natriuretic hormone (ANH), cardionatrine, cardiodilatine (CDD) and atriopeptin	These all are vasodilators and protein polypeptide hormones secreted by heart muscle cells. It is involved in the homeostatic control of body water, sodium, potassium and fat (adipose tissue).	Manufacturers of ANP: GL Biochem; Shanghai Hanhong Chemical Co.; Chemsky International Co.; Cellmano Biotech; LiuShi Pharmaceutical Technology

- Hairy cell leukemia.
- Genital warts.
- Diabetes.
- Chemotherapy-induced neutropenia.
- Autologous bone marrow transplantation.
- Anemia in zidovudine-treated HIV-infected patients.
- AIDS-related Kaposi's sarcoma.

References

[1] Romano P, Fleet G and Boekhout T 2008 Alcoholic fermentation: beverages to biofuel *FEMS Yeast Res.* **8** 965–6

[2] Van Mellaert L and Anné J 2001 Gram-positive bacteria as host cells for heterologous production of biopharmaceuticals *Novel Frontiers in the Production of Compounds for Biomedical Use. Focus on Biotechnology* vol 1 ed A Van Broekhoven, F Shapiro and J Anné (Dordrecht: Springer)

[3] Kamionka M 2011 Engineering of therapeutic proteins production in *Escherichia coli Curr. Pharm. Biotechnol.* **12** 268–74

[4] Nielsen J 2013 Production of biopharmaceutical proteins by yeast: advances through metabolic engineering *Bioengineered* **4** 207–11

[5] Ballard B E 1968 Biopharmaceutical considerations in subcutaneous and intramuscular drug administration *J. Pharm. Sci.* **57** 357–78

[6] Mitragotri S, Burke P A and Langer R 2014 Overcoming the challenges in administering biopharmaceuticals: formulation and delivery strategies *Nat. Rev. Drug Discov.* **13** 655–72

[7] Grenha A 2012 A systemic delivery of biopharmaceuticals: parenteral forever? *J. Pharm. Bioallied Sci.* **4** 95

[8] Zurdo J *et al* 2015 Early implementation of QbD in biopharmaceutical development: a practical example *BioMed. Res. Int.* **2015** 605427

[9] Spada S and Walsh G 2005 *Directory of Approved Biopharmaceutical Products* (Boca Raton, FL: CRC Press)

[10] Ohtori S *et al* 1976 Epidural administration of spinal nerves with the tumor necrosis factor-α inhibitor, etanercept, compared with dexamethasone for treatment of sciatica in patients with lumbar spinal stenosis: a prospective randomized study *Spine* **37** 439–44

[11] Wu-Pong S and Rojanasaku Y 2010 *Biopharmaceutical Drug Design and Development* (New York: Springer) p 205

[12] Compère V, Li S, Leprince J, Tonon M C, Vaudry H and Pelletier G 2003 Effect of intracerebroventricular administration of the octadecaneuropeptide on the expression of pro-opiomelanocortin, neuropeptide Y and corticotropin-releasing hormone mRNAs in rat hypothalamus *J. Neuroendocrinol.* **15** 197–203

[13] Majewska-Szczepanik M, Góralska M, Marcińska K, Zemelka-Wiącek M, Strzępa A, Dorożyńska I and Szczepanik M 2012 Epicutaneous immunization with protein antigen TNP-Ig alleviates TNBS-induced colitis in mice *Pharmacol. Rep.* **64** 1497–504

[14] Järvinen P A, Pennanen S and Ylöstalo P 1973 Induction of abortion by intra- and extra-amniotic prostaglandin F2 administration *Prostaglandins* **3** 491–504

[15] Doughty J C, Anderson J H, Willmott N and McArdle C S 1995 Intra-arterial administration of adriamycin-loaded albumin microspheres for locally advanced breast cancer *Postgrad. Med. J.* **71** 47–9

[16] Yovandich J, O'Malley B W, Sikes M and Ledley F D 1996 Gene transfer to synovial cells by intraarticular administrationof plasmid DNA *Hum. Gene Ther.* **39** 820–8

[17] Hengge U R, Walker P S and Vogel J C 1996 Expression of naked DNA in human, pig, and mouse skin *J. Clin. Invest.* **97** 2911–6

[18] RegeneRx Biopharmaceuticals 2015 *Clinical trial demonstrates elevated thymosin β4 plasma levels are associated with improvement of symptoms after stem cell therapy in patients with ischemic heart failure* CISION PR Newswire

[19] Garcia M S, Ono Y, Martinez S R, Chen S L, Goodarzi H, Phan T, Wehrli L N, Miyamura Y, Fung M A and Maverakis E 2011 Complete regression of subcutaneous and cutaneous metastatic melanoma with high-dose intralesional interleukin 2 in combination with topical imiquimod and retinoid cream *Melanoma Res.* **21** 235–43

[20] Fagan E A *et al* 1989 Adoptive immunotherapy administered via the hepatic artery and intralesional interleukin-2 in hepatocellular carcinoma *Cancer Treat Rev.* **16** 151–60

[21] Neves J and Sarmento B 2014 *Mucosal Delivery of Biopharmaceuticals: Biology, Challenges and Strategies* (New York: Springer) p 237

[22] Alawi K A, Morrison G C, Fraser D D, Al-Farsi S, Collier C and Kornecki A 2008 Insulin infusion via an intraosseous needle in diabetic ketoacidosis *Anaesth. Intensive Care* **36** 110–2

[23] Stuart G C, Nation J G, Snider D D and Thunberg P 1993 Intraperitoneal interferon in the management of malignant ascites *Cancer* **71** 2027–30

[24] Khabele D, Runowicz C D, Fields A L, Anderson P S and Goldberg G L 2003 Evaluation of low-dose intraperitoneal interferon-α for palliation of ascites in patients with non-ovarian gynecologic malignancies *Gynecol. Oncol.* **89** 420–3

[25] Acién P, Quereda F, Campos A, Gomez-Torres M J, Velasco I and Gutierrez M 2002 Use of intraperitoneal interferon *α*-2b therapy after conservative surgery for endometriosis and postoperative medical treatment with depot gonadotropin-releasing hormone analog: a randomized clinical trial *Fertil. Steril.* **78** 705–11

[26] Modarres M, Falavarjani K G, Nazari H, Sanjari M S, Aghamohammadi F, Homaii M and Samiy N 2011 Intravitreal erythropoietin injection for the treatment of non-arteritic anterior ischaemic optic neuropathy *Br. J. Ophthalmol.* **95** 992–5

[27] Li W, Sinclair S H and Xu G T 2010 Effects of intravitreal erythropoietin therapy for patients with chronic and progressive diabetic macular edema *Ophthalmic Surg. Lasers Imaging* **41** 18–25

[28] Kinnunen H M and Mrsny R J 2014 Improving the outcomes of biopharmaceutical delivery via the subcutaneous route by understanding the chemical, physical and physiological properties of the subcutaneous injection site *J. Control. Release* **182** 22–32

[29] Ho R J Y 2013 *Biotechnology and Biopharmaceuticals: Transforming Proteins and Genes into Drug* 2nd edn (New York: Wiley) p 681

[30] Choo Y C, Hsu C, Seto W H, Miller D G, Merigan T C, Ng M H and Ma H K 1985 Intravaginal application of leukocyte interferon gel in the treatment of cervical intraepithelial neoplasia (CIN) *Arch. Gynecol.* **237** 51–4

[31] Ramanathan R, Park J, Hughes S M, Lykins W R, Bennett H R and Hladik Woodrow K A 2015 Effect of mucosal cytokine administration on selective expansion of vaginal dendritic cells to support nanoparticle transport *Am. J. Reprod. Immunol.* **74** 333–44

[32] Jackson R L 2005 Biotechnology and biopharmaceuticals-transforming proteins and genes into drugs *J. Natl Med. Assoc.* **97** 1575

[33] Vela-Ojeda J, Tripp-Villanueva F, Sanchez-Cortés E, Ayala-Sanchez M, García-Ruiz Esparza M A, Rosas-Cabral A and Gonzalez-Llaven J 1999 Intravesical rhGM-CSF for the treatment of late onset hemorrhagic cystitis after bone marrow transplant *Bone Marrow Transplant.* **24** 1307–10

[34] Lee J, Jung E, Park J and Park D 2005 Transdermal delivery of interferon-gamma (IFN-gamma) mediated by penetratin, a cell-permeable peptide *Biotechnol. Appl. Biochem.* **42** 169–73

[35] Karran E and Hardy J 2014 Antiamyloid therapy for Alzheimer's disease—are we on the right road? *N. Engl. J. Med.* **370** 377–8

[36] Konsman J P, Tridon V and Dantzer R 2000 Diffusion and action of intracerebroventricularly injected interleukin-1 in the CNS *Neuroscience* **101** 957–67

[37] Pawliński R and Janeczko K 1997 Intracerebral injection of interferon-gamma inhibits the astrocyte proliferation following the brain injury in the 6-day-old rat *J. Neurosci. Res.* **50** 1018–22

[38] Bayley D, Temple C, Clay V, Steward A and Lowther N 1995 The transmucosal absorption of recombinant human interferon-α B/D hybrid in the rat and rabbit *J. Pharm. Pharmacol.* **47** 721–4

[39] Walsh G 2004 Second-generation biopharmaceuticals *Eur. J. Pharmaceut. Biopharmaceut.* **58** 185–96

[40] Walsh G 2010 Post-translational modifications of protein biopharmaceuticals *Drug Discov. Today* **15** 773–80

[41] Guiochon G and Beaver L A 2011 Separation science is the key to successful biopharmaceuticals *J. Chromatogr. A* **1218** 8836–58

[42] Crommelin D J A, Storm G, Verrijk R, de Leede L, Jiskoot W and Hennink W E 2003 Shifting paradigms: biopharmaceuticals versus low molecular weight drugs *Int. J. Pharm.* **266** 3–16

[43] Doulatov S, Notta F, Laurenti E and Dick J E 2012 Hematopoiesis: a human perspective *Stem Cell* **10** 120–36

[44] Vilcek J and Feldmann M 2004 Historical review: cytokines as therapeutics and targets of therapeutics *Trends Pharmacol. Sci.* **25** 201–9

[45] Pluda J M, Yarchoan R, Smith P D, McAtee N, Shay L E, Oette D, Maha M, Wahl S M, Myers C E and Broder S 1990 Subcutaneous recombinant granulocyte-macrophage colony-stimulating factor used as a single agent and in an alternating regimen with azidothymidine in leukopenic patients with severe human immunodeficiency virus infection *Blood* **76** 463–72

[46] Clarke V, Dunstan F D and Webb D K 1999 Granulocyte colony-stimulating factor ameliorates toxicity of intensification chemotherapy for acute lymphoblastic leukemia *Med. Pediatr. Oncol.* **32** 331–5

[47] Paul M *et al* 2014 Subcutaneous versus intravenous granulocyte colony stimulating factor for the treatment of neutropenia in hospitalized hemato-oncological patients: randomized controlled trial *Am. J. Hematol.* **89** 243–8

[48] Honkoop A H, Hoekman K, Wagstaff J, van Groeningen C J, Vermorken J B, Boven E and Pinedo H M 1996 Continuous infusion or subcutaneous injection of granulocyte-macrophage colony-stimulating factor: increased efficacy and reduced toxicity when given subcutaneously *Br. J. Cancer* **74** 1132–6

[49] Bhadra D, Bhadra S, Jain P and Jain N K 2002 Pegnology: a review of PEG-ylated systems *Pharmazie* **57** 5–29
[50] Amouzadeh H R, Engwall M J and Vargas H M 2015 Safety pharmacology evaluation of biopharmaceuticals *Handb. Exp. Pharmacol.* **229** 385–404
[51] Serabian M A and Pilaro A M 1999 Safety assessment of biotechnology-derived pharmaceuticals: ICH and beyond *Toxicol. Pathol.* **27** 27–31
[52] Baumann A 2009 Nonclinical development of biopharmaceuticals *Drug Discov. Today* **14** 1112–22
[53] Richter W F, Bhansali S G and Morris M E 2012 Mechanistic determinants of biotherapeutics absorption following SC administration *AAPS J.* **14** 559–70
[54] McDonald T A, Zepeda M L, Tomlinson M J, Bee W H and Ivens I A 2010 Subcutaneous administration of biotherapeutics: current experience in animal models *Curr. Opin. Mol. Ther.* **12** 461–70
[55] Richter W F and Jacobsen B 2014 Subcutaneous absorption of biotherapeutics: knowns and unknowns *Drug Metab. Dispos.* **42** 1881–9
[56] Vugmeyster Y, Xu X, Theil F P, Khawli L A and Leach M W 2012 Pharmacokinetics and toxicology of therapeutic proteins: advances and challenges *World J. Biol. Chem.* **3** 73–92
[57] Basu S K, Govardhan C P, Jung C W and Margolin A L 2004 Protein crystals for the delivery of biopharmaceuticals *Expert Opin. Biol. Ther.* **4** 301–17
[58] Mengiardi S, Tsakiris D, Molnar V, Kohlhaas-Styk U, Mittag M, Kraehenbuehl S and Hersberger K 2014 Impact of pharmaceutical care on self-administration of outpatient low-molecular-weight heparin therapy *Pharm. Pharm.* **5** 372–85
[59] Gundersen L E 2001 Training needs in regulatory science for the biopharmaceutical industry *Nat. Biotechnol.* **19** 1187–8
[60] Engle J P, Kraus D M, Parent L S and Dean-Holland M 1993 The pharmacist practitioner's role in biotechnology: clinical application of biotechnology products *Biotechnology and Pharmacy* ed J M Pezzuto, M E Johnson and H R Manasse (Dordrecht: Springer)
[61] Frederiksen B, Pressler T, Hansen A, Koch C and Høiby N 2006 Effect of aerosolized rhDNase (Pulmozyme) on pulmonary colonization in patients with cystic fibrosis *Acta Paediatr.* **95** 1070–4
[62] Boe J *et al* 2001 European respiratory society guidelines on the use of nebulizers: guidelines prepared by a European respiratory society task force on the use of nebulizers *Eur. Respir. J.* **18** 228–42
[63] Pressler T 2008 Review of recombinant human deoxyribonuclease (rhDNase) in the management of patients with cystic fibrosis *Biologics* **2** 611–7
[64] Khan S *et al* 2016 Role of recombinant DNA technology to improve life *Int. J. Genomics* **2016** 2405954
[65] Giezen T J, Mantel-Teeuwisse A K and Leufkens H G 2009 Pharmacovigilance of biopharmaceuticals: challenges remain *Drug Saf.* **32** 811–7
[66] Hörl W H 1999 Is there a role for adjuvant therapy in patients being treated with epoetin? *Nephrol. Dial. Transplant.* **14** 50–60
[67] Evans H P, Cooper A, Williams H and Carson-Stevens A 2016 Improving the safety of vaccine delivery *Hum. Vaccin. Immunother.* **12** 1280–1
[68] Casella S J, Mongilio M K, Plotnick L P, Hesterberg M P and Long C A 1993 Accuracy and precision of low-dose insulin administration *Pediatrics* **91** 1155–7

[69] Jefferis R 2013 Review of glycosylation engineering of biopharmaceuticals: methods and protocols: a book edited by Alain Beck *MAbs* **5** 638–40

[70] Wilson A W and Neumann P J 2012 The cost-effectiveness of biopharmaceuticals *MAbs* **4** 281–8

[71] Nash D B 2012 The use of medicines in the United States: a detailed review *Am. Health Drug Benefits* **5** 423

[72] Moorkens E, Meuwissen N, Huys I, Declerck P, Vulto A G and Simoens S 2017 The market of biopharmaceutical medicines: a snapshot of a diverse industrial landscape *Front. Pharmacol.* **8** 314

[73] Blackstone E A and Fuhr J P 2007 Generic biopharmaceutical drugs: an economic and policy analysis *Biotechnol. Healthc.* **4** 43–8

[74] Devine J W, Cline R R and Farley J F 2006 Follow-on biologics: competition in the biopharmaceutical marketplace *J. Am. Pharm. Assoc.* **46** 193–201

[75] Simoens S 2009 Health economics of market access for biopharmaceuticals and biosimilars *J. Med. Econ.* **12** 211–8

[76] Simoens S 2011 Biosimilar medicines and cost-effectiveness *Clinicoecon. Outcomes Res.* **3** 29–36

[77] Blackstone E A and Fuhr Jr J P 2012 Innovation and competition: will biosimilars succeed? *Biotechnol. Healthc.* **9** 24–7

[78] Poirier S, Buffington D E and Memoli G A 1999 Billing third party payers for pharmaceutical care services *J. Am. Pharm. Assoc.* **39** 50–64

[79] McElhiney L F 2013 Optimizing third-party reimbursement for compounded medications *Int. J. Pharm. Compd.* **17** 201–4

[80] Ganther J M 2002 Third party reimbursement for pharmacist services: why has it been so difficult to obtain and is it really the answer for pharmacy? *J. Am. Pharm. Assoc.* **42** 875–9

[81] Nold E G and Pathak D S 1977 Third-party reimbursement for clinical pharmacy services: philosophy and practice *Am. J. Hosp. Pharm.* **34** 823–6

[82] Fugel H J, Nuijten M and Postma M 2012 Stratified medicine and reimbursement issues *Front. Pharmacol.* **3** 181

[83] Ganther J M and Kreling D H 1999 Effect of a change in third party reimbursement rate on prescription gross margin *J. Am. Pharm. Assoc.* **39** 346–52

[84] Akaho E, MacLaughlin E J and Takeuchi Y 2003 Comparison of prescription reimbursement methodologies in Japan and the United States *J. Am. Pharm. Assoc.* **43** 519–26

[85] Park M Y and Ahnb K Y 2013 Effect of the refrigerator storage time on the potency of botox for human extensor digitorum brevis muscle paralysis *J. Clin. Neurol.* **9** 157–64

[86] Kleinberg M L 1995 Sterility of repackaged filgrastim and sargramostim *Am. J. Health Syst. Pharm.* **52** 1101

[87] Plews-Ogan M L *et al* 2004 Patient safety in the ambulatory setting: a clinician-based approach *J. Gen. Intern. Med.* **19** 719–25

[88] Orive G, Gascón A R, Hernández R M, Domínguez-Gil A and Pedraz J L 2004 Techniques: new approaches to the delivery of biopharmaceuticals *Trends Pharmacol. Sci.* **25** 382–7

[89] Lins L and Brasseur R 1995 The hydrophobic effect in protein folding *FASEB J.* **9** 535–40

[90] Wang W 1999 Instability, stabilization, and formulation of liquid protein pharmaceuticals *Int. J. Pharmaceut.* **185** 129–88

[91] Chang L L and Pikal M J 2009 Mechanisms of protein stabilization in the solid state *J. Pharm. Sci.* **98** 2886–908

[92] Jacob S 2006 Stablity of proteins in aqueous solution and solid state *Indian. J. Pharmaceut. Sci.* 154–63

[93] Manning M C, Patel K and Borchardt R T 1989 Stability of protein pharmaceuticals *Pharmaceut. Res.* **6** 903–18

[94] Berkowitz S A, Engen J R, Mazzeo J R and Jones G B 2012 Analytical tools for characterizing biopharmaceuticals and the implications for biosimilars *Nat. Rev. Drug Discov.* **11** 527–40

[95] Bhatia S 2018 *Introduction to Pharmaceutical Biotechnology: Enzymes, Proteins and Bioinformatics* vol 2 (Bristol: IOP Publishing)

[96] Bhatia S 2018 *Introduction to Pharmaceutical Biotechnology: Basic Techniques and Concepts* vol 1 (Bristol: IOP Publishing)

Lightning Source UK Ltd.
Milton Keynes UK
UKHW050638100419
340794UK00003BA/42/P